The Changing Economic Geography of Companies and Regions in Times of Risk, Uncertainty, and Crisis

This book offers conceptual and empirical insights from economic geography to explore how uncertainties, crises, and risks, shape, reshape, and ultimately transform the spatial arrangements of companies and regions.

This book provides valuable insights into the evolving landscape of economic interactions amidst contemporary challenges. It explores concepts such as global value chains, global production networks, regional resilience, and the impact of crises, risks, and uncertainties on spatial economic patterns. Case studies from various regions, nations, and industries, including lesser researched sectors such as medical technology and the restaurant and bar industry offer tangible and real-world manifestations of these dynamics. Through its comprehensive coverage and interdisciplinary approach, this book equips readers with practical knowledge applicable to academia and real-world contexts. It offers a deeper understanding of the complex interplay between economic geography and contemporary challenges at a variety of levels.

The book provides valuable perspectives for academics, practitioners, and policy makers in the fields of Economic Geography, Regional Studies, Political Sciences, Economic Sociology, Economics, and International Business Studies.

Thomas Neise is an Interim Professor of Economic and Social Geography at the Heidelberg University in Germany. His research focuses on the adaptation and resilience strategies of companies to extreme weather events, climate change, and crises. Another research interest lies on risks in global production networks. He also studies the role of foreign direct investment for regional development in Southeast Asia.

Philip Verfürth is a post-doctoral researcher at the Institute of Geography, Osnabrück University. He has been a visiting scholar at Ankara University and the Leibniz Institute for Research on Society and Space in Erkner, Germany. His research interests include the spatial dynamics of globalization, with a focus on global production networks, the geographies of digital transformation, and the intersection of firm resilience and regional development.

Martin Franz holds the Chair for Human Geography with an emphasis on Economic Geography at Osnabrück University. After studying geography in Bochum, Martin Franz worked at the Center for Interdisciplinary Ruhr Research (ZEFIR) in Bochum and the Departments of Geography at the Philipps-University of Marburg and the University of Bayreuth.

The Dynamics of Economic Space

This series aims to play a leading international role in the development, promulgation and dissemination of new ideas in economic geography. It has as its goal the development of a strong analytical perspective on the processes, problems and policies associated with the dynamics of local and regional economies as they are incorporated into the globalizing world economy. In recognition of the increasing complexity of the world economy, the Commission's interests include: industrial production; business, professional and financial services, and the broader service economy including e-business; corporations, corporate power, enterprise and entrepreneurship; the changing world of work and intensifying economic interconnectedness.

Social Protection and Informal Workers in Sub-Saharan Africa
Lived Realities and Associational Experiences from Tanzania and Kenya
Edited by Lone Riisgaard, Winnie Mitullah, and Nina Torm

Economies, Institutions and Territories
Dissecting Nexuses in a Changing World
Edited by Luca Storti, Giulia Urso and Neil Reid

Excessive Inequality and Socio-Economic Progress
Ona Gražina Rakauskienė, Dalia Streimikiene and Lina Volodzkienė

Blue Economy
People and Regions in Transitions
Edited by C. Patrick Heidkamp, John Morrissey and Celine Germond-Duret

Inequalities, Youth and the Labour Market
NEETS in Southern Europe
Edited by Theodoros Iosifides, Stelios Gialis and Thanasis Kizos

The Changing Economic Geography of Companies and Regions in Times of Risk, Uncertainty, and Crisis
Edited by Thomas Neise, Philip Verfürth, and Martin Franz

The Changing Economic Geography of Companies and Regions in Times of Risk, Uncertainty, and Crisis

Edited by Thomas Neise, Philip Verfürth, and Martin Franz

Routledge
Taylor & Francis Group
LONDON AND NEW YORK

First published 2025
by Routledge
4 Park Square, Milton Park, Abingdon, Oxon OX14 4RN

and by Routledge
605 Third Avenue, New York, NY 10158

Routledge is an imprint of the Taylor & Francis Group, an informa business

British Library Cataloguing-in-Publication Data
A catalogue record for this book is available from the British Library

ISBN: 978-1-032-60107-6 (hbk)
ISBN: 978-1-032-60109-0 (pbk)
ISBN: 978-1-003-45762-6 (ebk)

DOI: 10.4324/9781003457626

Typeset in Times New Roman
by SPi Technologies India Pvt Ltd (Straive)

Contents

Figures

Tables

Contributors

John R. Bryson has held research and teaching positions at the Universities of St Andrews, Cambridge, Wales, and most recently at the University of Birmingham where he is Professor of Enterprise and Economic Geography. His research focuses on understanding people and organizations and place-based adaptation. His books include Service Business Theory and Practice (2020), Ordinary Cities, Extraordinary Geographies: People, Place and Space (2021), and Pandemic Recovery? Reframing and Rescaling Societal Challenges (2024).

Felix Bücken is a research associate at the Institute of Geography at Osnabrück University. His research interests include the resilience of companies to climate risks and the role of regional networks for organizational adaptation. Prior to his academic work, he worked as a business developer in areas such as the promotion of key industries and the design of programs to secure skilled labor.

Han Chu is a doctoral student and research associate at the Working Group Economic Geography at Kiel University in Germany. In 2020, she received a Master in Urban Planning from Guangdong University of Technology, China. Her research focuses on digital platforms, regional and cluster development, and regional resilience.

Przemysław Ciesiółka holds a Ph.D. in Geography and a M.Sc. in Planning. He is an assistant professor at the Adam Mickiewicz University in Poznań, Poland. He has studied and completed internships at the University of Portsmouth, Christian-Albrecht University in Kiel, and Comenius University in Bratislava. Mr. Ciesiółka is a national representative in the Association of European Schools of Planning and a specialist in the planning and management of local government units, with special emphasis on regeneration.

Hilal Erkuş is a full professor at Akdeniz University (Faculty of Architecture, Department of Urban and Regional Planning, Antalya, Turkey). She is specialized on urban and regional economic development, (political) economic (tourism) geography, tourism clusters networks, tourism value chains, urban tourism governance, economic diversification in tourism, survival of tourism enterprises, and fictional expectations in tourism.

Martin Franz holds the Chair for Human Geography with an emphasis on Economic Geography at Osnabrück University. After studying geography in Bochum, Martin Franz worked at the Center for Interdisciplinary Ruhr Research (ZEFIR) in Bochum and the Departments of Geography at the Philipps-University of Marburg and the University of Bayreuth.

Robert Hassink is a Professor of Economic Geography at Kiel University in Germany and a Visiting Professor at Newcastle University, UK. Since his Ph.D. in 1992, he has carried out various research projects on old industrial areas, clusters, regional innovation policy, and border regions in Europe and East Asia, as well as on theories and paradigms of economic geography.

Frank Jicha is a funding consultant and project manager at the Business Development Agency Kreis Plön GmbH. He worked as a research associate at the Working Group Economic Geography at Kiel University in Germany from January 2022 to September 2023.

Ceyda Kurtar Anlı is a research assistant in the Department of Geography at Ankara University, Turkey. She holds a Ph.D. degree and has completed both a Master's and Ph.D. degrees at the same institution. Her research interests focus on subfields of economic geography, including consumption geography, retail geography, and industrial clustering.

Michał Męczyński holds a Ph.D. in Geography and a M.Sc. in Planning. Mr. Męczyński is an Associate Professor at the Adam Mickiewicz University in Poznań, Poland. He completed internships at, e.g., European University Viadrina Frankfurt (Oder), the University of Birmingham, and Christian-Albrecht University of Kiel. His research interests focus on the problems of spatial management and economic geography, especially spatial differences in the creativity development and innovativeness of business entities. His most recent interests navigate around the problem of transformation of old industrial areas based on the eco-innovativeness of local enterprises.

Thomas Neise is an Interim Professor of Economic and Social Geography at the Heidelberg University in Germany. His research focuses on the adaptation and resilience strategies of companies to extreme weather events, climate change, and crises. Another research interest lies on risks in global production networks. He also studies the role of foreign direct investment for regional development in Southeast Asia.

Piotr Pachura is currently a University Professor at the Management Faculty of Częstochowa University of Technology (Poland), Head of Department of Management and Entrepreneurship. He holds academic degrees in social science (M.Sc.), economic science (Ph.D.), and economic geography (post-doc).

His main research field is the interdisciplinary interplay between "spaces – people – organizations – institutions".

Mehmet Ragıp Kalelioğlu is currently working as an Assistant Professor at the Department of Geography, Gaziantep University, Turkey. His research interests concentrate on the subfields of Economic Geography such as retail geography, industrial clustering, innovation, blue economy, and urban economy. Dr. Kalelioğlu took part as a researcher in two prestigious projects on innovation and foreign direct investment in the field of regional development.

Pieter Terhorst is a Professor Emeritus at the University of Amsterdam (Department of Planning, Geography and International Development Studies, Amsterdam, The Netherlands.) He is specialized on political economic geography, variety of capitalism, tourism cities, clusters, rescaling of state spaces, and fictional expectations.

Philip Verfürth is a post-doctoral researcher at the Institute of Geography, Osnabrück University. He has been a visiting scholar at Ankara University and the Leibniz Institute for Research on Society and Space in Erkner, Germany. His research interests include the spatial dynamics of globalization, with a focus on global production networks, the geographies of digital transformation, and the intersection of firm resilience and regional development.

Philip Völlers received his Ph.D. in Geography from Osnabrück University in Germany and holds a master's degree in Economic and Social Geography. His research is focused on examining the varying effects of risk-related processes on the strategic responses of both firm and non-firm actors in global production networks. His area of expertise is the Turkish–German bilateral direct investment context.

Nuri Yavan is a full professor at the Department of Geography at Ankara University, Turkey. His research interests and publications focus on three areas: Economic Geography, particularly industrial and FDI location; Regional Development and Policy, especially innovation and institutions; and History and Philosophy of Geography. He has published 9 books, 23 book chapters, 25 research/project reports, and 43 articles in academic journals and presented more than 140 papers at national and international conferences. He served as the Chair of the IGU Commission on the Dynamics of Economic Spaces.

Acknowledgments

We would like to thank a number of people for their help in preparing and finalizing this volume. Nuri Yavan, in his function as the Chair of the IGU Commission on the Dynamics of Economic Spaces, provided invaluable insights and support to set up this volume. Josephine Bruchmann helped us with her conscientious work in formatting and editing the contributions of this volume. We would also like to express our appreciation to our reviewers Crispian Fuller, Sina Hardaker, Philip Völlers, and Nuri Yavan, who have helped to improve the content of the chapters. Prachi Priyanka's support on the publisher's side was instrumental in navigating the publication process smoothly. We would also like to thank all participants from the IGU-CDES conference "Reshaping Economic Geography in Times of Risk, Uncertainty and Crisis", which took place in October 2022 in Osnabrück, Germany and on whose selected papers this volume is based.

1 Introduction

Changing Economic Geographies in Times of Risk, Uncertainty, and Crisis

Thomas Neise, Martin Franz, and Philip Verfürth

1.1 Setting the Stage

The first half of the 2020s was and is characterized by overlapping political, environmental, social, and economic crises, which have mutually reinforced each other and have had, and continue to have, strong spatial impacts. The predominant crisis that has affected social and economic relations since the beginning of the 2020s, and ultimately also influenced economic geography debates, is the COVID-19 pandemic. The pandemic led, for instance, to disruptions in supply chains, reduced consumer demand, business closures, and changes in consumer behavior. Governments around the world implemented various measures such as lockdowns and travel restrictions to contain the virus, contributing to a global economic downturn (e.g., Onyeaka et al. 2021). After the pandemic entered its final phase, Russia's war of aggression against Ukraine added a further crisis that is having a massive impact on the global economy through financial and trade sanctions, increasing commodity prices, disruptions to supply chains, and rising expenditure on armaments (e.g., Liadze et al. 2023). In addition to these events, the effects of climate change continue, leading to more frequent and severe weather events that damage infrastructure, including roads, bridges, and buildings, resulting in more disruptions of supply chains and significant economic costs for repairs and reconstruction (e.g., Kalkuhl and Wenz 2020). These are just three of the biggest and most global crises with large spatio-economic impacts; however, many other developments have also led to uncertainties and risks for companies and regions. Although we have outlined these three moments of crisis separately, in fact, they are interconnected and overlap spatially. Even more, these crises generate a synergistic effect that results in more extensive damage than the combined impact of individual crises: what is referred to as a "poly-crisis" (see Henig and Knight 2023; Lawrence, Janzwood, and Homer-Dixon 2022; Tooze 2022).

Due to the great significance of crises, and the risks and uncertainties that underlie them or emanate from them, they require an intensive scientific examination, or as Brinks and Ibert (2020, 276) put it: "If we accept the diagnosis that we live in times of crises, it becomes more important than ever to develop

DOI: 10.4324/9781003457626-1

a more profound understanding of crisis as a particular context for action and a possibility for intervention." The same is true for uncertainties and risks.

The concept of uncertainty has long been a subject of discussion across various fields, sparking ongoing debates regarding its definition and comprehension. Knight's (1921) distinction between risk and uncertainty in the economic context is a crucial contribution. He describes risks as situations where probabilities can be assigned, allowing for ex ante calculations and strategic decision-making based on past statistics. Conversely, uncertainty pertains to situations without reliable probability calculations, often associated with unique circumstances, major changes, innovations, and crises (Beckert 1996). Knight's (1921) objectifying perspective on risk, still finds predominant application in economic geographical discussions. These discussions often revolve around a risk-and-reward paradigm or a trade-off mindset, considering the interplay between risk and opportunities (e.g., Bathelt and Henn 2014; Blažek 2016; Bryson and Vanchan 2020).

Knight's (1921) analytical separation of uncertainty and risk is clear, but determining whether a situation is uncertain or risky remains difficult. Even seemingly predictable situations with probability distributions contain elements of uncertainty that limit the possibilities of rational economic calculation. In economic geography literature, some scholars advocate for an actor-centered approach, with a focus on socio-relational and socio-spatial aspects. In this context, they highlight variables such as spatial–relational context (e.g., Ekinsmyth 2002; November 2008; Rodrigue, Notteboom, and Pallis 2011), emphasizing embeddedness and the role of risk perception (e.g., Völlers, Yavan, and Franz 2021; Völlers et al. 2023). These discussions contribute to alternative conceptual arguments regarding the nature of risk, portraying it as a product of social practices rather than an objective entity (e.g., Brill and Robin 2020; Schwabe 2020; Völlers et al. 2023).

How individuals and organizations react to uncertainties, risks, and crises depends very much on perceptions and discourses. This can also be seen in the examples mentioned above. Reactions to the COVID-19 pandemic (e.g., Balmford et al. 2020), the war in Ukraine (e.g., Mol, Rabbiosi, and Santangelo 2023) and climate change (e.g., Skjærseth, Bang, and Schreurs 2013) not only differ greatly between countries or regions, but there are also major differences between companies or individuals in terms of how strongly they perceive threats to their own way of life, way of working or business model. These differences in perception result in different reactions and strategies for dealing with the uncertainties and risks associated with these crises. This can range from refusing to recognize the existence of a (potential) problem and possibly even circumventing state-prescribed crisis management measures (e.g., pandemic measures), to proactively implementing adaptation or resilience strategies (e.g., Neise, López, and Angga Reksa 2023).

The examples mentioned at the beginning of this introduction already highlight diseases, geopolitical tensions, and human-made environmental catastrophes as possible sources of uncertainties and risks, but technological and societal changes, instability of organizations such as states, or natural disasters like earthquakes may also be sources. These uncertainties and risks can manifest especially in crises, where the normal functioning of economic systems is disrupted, often leading to severe consequences for individuals and communities. Crises are defined as "a serious threat to the basic structures or the fundamental values and norms of a system, which under time pressure and highly uncertain circumstances necessitates making critical decisions" (Rosenthal et al. 1989, 10). Brinks and Ibert (2020, 10), sum up that a crisis "encompasses the elements of uncertainty, urgency and threat."

Adapting to uncertainties, risks, and crises through resilience strategies is a crucial factor in avoiding and managing crises. In its original understanding, resilience is understood as a simple bounce back from shocks to the original situation. This so-called engineering resilience perspective "is grounded on a machine-like view of systems, with simple cause and effect relationships" (Barasa, Mbau, and Gilson 2018, 496). In this time of crisis, we can observe how crises pushed new ways of economic exchange and strategies. As a result, individuals, organizations, sectors, and institutions at the regional, national, and global levels are and will be confronted with the need to respond to these uncertainties and dynamics.

Companies have developed diverse strategies and responses to address crises, risks, and uncertainties, depending on their industry, size, and the specific nature of the challenge. Many organizations have implemented and expanded crisis management planning and risk mitigation strategies. Supply chain diversification has become a significant focus, although it is often balanced against cost-saving considerations. In response to crises such as the COVID-19 pandemic, numerous companies embraced remote work and expedited their digital transformation efforts (see e.g., Haraguchi et al. 2023). Crises have not only stimulated innovation in the realm of digital transformation (e.g., Buck et al. 2022) but also in other areas, as companies actively seek novel solutions and opportunities (e.g., Ebersberger and Kuckertz 2021). Cost-cutting measures and a reevaluation of financial strategies have been common responses to crises and risks (e.g., Van Egdom et al. 2022), leading to the postponement of investments in many cases (e.g., Buchheim et al. 2020). Consumer behavior changes during crises have forced companies to make adjustments to meet evolving demands (Franz, Neise, and Verfürth 2023). Overall, companies are navigating uncertainties with adaptability and strategic responses tailored to the unique circumstances of each crisis.

In the face of these reactions to the multiple contemporary crises, the concept of resilience has gained increasing prominence in economic geography, leading to a greater variety of resilience concepts and understandings. A valuable contribution to the literature on regional resilience is the emergence of transformative resilience, which emphasizes a region's capacity to steer toward sustainability (Martin and Sunley 2020; Trippl, Fastenrath, and Isaksen 2023). Furthermore, scholars are engaged in analyzing resilience within regions and sectors in the wake of crises, using terms such as regional resilience (e.g., Gong et al. 2020;

Lüder and Kalvelage, 2023), business resilience (e.g., Dahles and Susilowati 2015; Soroka et al. 2020), and organizational resilience (e.g., Barasa et al., 2018; Neise, Sambodo, and Revilla Diez 2021). All of this research highlights that local characteristics play an important role in achieving resilience during or after a crisis, emphasizing that resilience depends on specific contextual factors and the actions of individuals (Bristow and Healy 2014; David 2018; Kitsos and Bishop 2018). These are just examples of resilience concepts that have emerged in economic geography. But how can existing theories and concepts in economic geography provide answers for this contemporary and future world, and where is there a need to rethink economic geography research?

This book addresses these issues by providing conceptual and empirical contributions on how spatial configurations are influenced, reshaped, and eventually transformed by uncertainties, crises, and risks. Therefore, the book aims to explore the "reshaping of economic geography" in two ways. First, the book seeks to provide theoretical debates on how contemporary concepts of economic geography, such as global value chains (GVC), global production networks (GPN), and regional resilience, can contribute to the understanding of how crises, risks, and uncertainties affect spatial economic patterns (see Bryson 2024; Jicha, Hassink, and Chu 2024; Terhorst and Erkuş 2024). Second, the book provides empirical case studies on how crises, risks, and uncertainties reshape spatial configurations (Męczyński and Ciesiółka 2024; Pachura 2024; Yavan et al. 2024). The empirical case studies focus on various regions and nations as well as a range of different business sectors, including cases from under-researched sectors, such as the medical technology and restaurant and bar industry (see Franz et al. 2024; Neise 2024).

The book presents selected presentations from the conference "Reshaping Economic Geography in Times of Risk, Uncertainty and Crisis" under the umbrella of the International Geographical Union (IGU) and its commission on "The Dynamics of Economic Spaces" (CDES) in Osnabrück, Germany in October 2022. The CDES is dedicated to advancing international research, scholarship, and collaboration in economic geography. The annual conferences provide opportunities for a more thorough discussion of a given topic in economic geography, and to deepen international relations in economic geography.

1.2 Structure of the Book

The book is organized into three distinctive sections, each offering profound insights into the intricate spatial dynamics of crises, risks, and uncertainties, the related response by companies, and the transformation of regional or national economies. Section A, consisting of four chapters, provides strong conceptually oriented chapters that draw on recent dominant debates on regional resilience, fictional expectations, GPN and GVC associated with risks, uncertainties, and crises. The two chapters in Section B present case studies that deal with how companies and regions achieve resilience in moments of crisis. The last Section C includes three chapters that focus on the response to risks and uncertainties by companies and regions exposed to contemporary challenges of the globalized economy.

1.2.1 Section A: Conceptual Discussion of Uncertainties, Risks, and Crises within Economic Geography

The journey begins with Chapter 2 by Verfürth et al. (2024) that provides an overview of the conceptualization and appearance of the terms: uncertainty, risk, and crisis in the economic geography literature. Drawing on several disciplinary perspectives (e.g., economics, sociology, organizational studies, and political science), Verfürth et al. (2024) discuss the conceptual nuances and theoretical foundations of uncertainties, risks, and crises. The authors conclude that distinctions exist in the economic geography literature regarding the scope and depth of the discussion of risks, crises, and uncertainties. While there is a robust theoretical and empirical discourse on risks and crises, uncertainty receives less conceptual and empirical attention in the economic geography debate.

Following this, Chapter 3 by Jicha, Hassink, and Chu (2024) advances the discourse on regional economic resilience. Based on the recent multiple crises that have affected regions worldwide, the authors emphasize that regional economies are interconnected through GVC. As the concept of regional resilience is less concerned with these extra-regional factors that influence the resilience of regional economies, Jicha, Hassink, and Chu (2024) propose the concept of regional value chain resilience. Incorporating insights from the GVC literature, the authors stress that regional economies need to find an optimal level of resilience and efficiency. As an outlook, the authors also derive several research questions for future research on the impact of GVC resilience on the geography of production.

The exploration of risks and uncertainties within GPN and GVC research takes an interesting turn with Chapter 4 by Bryson (2024). Bryson outlines the transition from the risk society to Jenga capitalism and places it within a multi-scalar framework characterized by the application of a micro-foundational perspective. He highlights that this new era of Jenga capitalism is associated with new types of risk as a result of increasing system convergence. Due to deepening globalization, Bryson (2024) argues for a focus on plexuses rather than simple network thinking, as this better represents the complexities of the globalized economy. With the example of cyber-energy-production plexuses, he makes a valuable contribution to the literature on risks in GVC and GPN in recognizing multiple risks that can lead to parallel shocks or shock chains.

Closing Section A, Chapter 5 by Terhorst and Erkuş (2024) offers an exploration into the cognitive aspects of economic challenges, connecting the concept of regional economic resilience with that of fictional expectations about the future. Terhorst and Erkuş (2024) illustrate their conceptual arguments on fictional expectations in regional economic resilience with two case studies. The first case study is about the Port Authority of Rotterdam and its response to future shocks due to sea level rise and geopolitical conflicts. The authors show that the Port Authority of Rotterdam operates with scenarios to implement anticipatory adaptation measures, such as reorienting or reorganizing its infrastructure investment planning. The second case study looks at the tourism industry in Antalya in the wake of the COVID-19 crisis. The chapter presents how local actors are advancing their diversification activities to make the local tourism economy more resilient.

1.2.2 Section B: Economic Transition and Resilience of Companies and Regions

The book then proceeds to Section B, where the focus is on the economic transition and resilience of companies and regions. Chapter 6 by Franz et al. (2024) looks at the impact of the COVID-19 pandemic on the restaurant and bar industry. As businesses have faced major difficulties during and after the pandemic, the chapter discusses how economic development and city marketing agencies can support the resilience of restaurants and bars. Based on the concept of organizational resilience, the authors show how businesses have coped with the consequences of the pandemic in the short term. In addition, the chapter emphasizes the importance of developing local concepts for the restaurant and bar industry and strengthening business-related expertise in economic development and city marketing agencies, as well as the need to improve networking and knowledge transfer between businesses and between city marketing and economic development agencies and the business owners.

Continuing this exploration, Chapter 7 by Męczyński and Ciesiółka (2024) examines how old industrial regions are trying to escape the crisis associated with fossil fuel dependency by promoting eco-innovations. The authors draw on the concept of lock-in and path dependency to elaborate on the challenges of regional green transformation processes. Using the Polish province Silesia, the chapter looks at the regional spatial differences of eco-innovations to present the status quo of the regional green transformation.

1.2.3 Section C: Uncertainties and Risks in the Global Economy

The final section of the book deals with the uncertainties and risks prevalent in the global economy. Chapter 8 by Neise (2024) contributes to the emerging GPN literature on risk. Applying the new concept of performative risk narrative to the case of the medical technology industry, the chapter analyzes how risk narratives influence the architecture and reconfiguration strategies of GPN. The author shows that risk narratives related to safety concerns and the dominant role of global suppliers, limit the ability of medical technology firms to create more flexible and reliable production networks in moments of crisis.

Chapter 9 by Yavan et al. (2024), scrutinizes the digital transformation in peripheral regions. Focusing on the carpet industry in Gaziantep, the authors reveal the risks that Gaziantep faces during the digitalization process, and how it is attempting to adapt to digital transformation to mitigate these risks. Therewith the study contributes to the economic geography literature on industry 4.0 and the spatiality of digitalization.

Chapter 10 by Pachura (2024), which concludes the book's diverse perspectives, looks at the impact of socio-cultural factors on the choice of location for foreign direct investment. Conceptually, the chapter draws on the literature on risk perception in relation to foreign investment locations. With a focus on the Polish city of Czestochowa, which is both an important destination for foreign direct investment and a global religious center, Pachura (2024) provides an interesting case study that shows how the local religious context can shape the risk perception of foreign investors.

References

Balmford, B., J. D. Annan, J. C. Hargreaves, M. Altoè, and I. J. Bateman. 2020. Cross-country comparisons of Covid-19: policy, politics and the price of life. *Environmental and Resource Economics* 76:525–551. doi: 10.1007/s10640-020-00466-5

Barasa, E., R. Mbau, and L. Gilson. 2018. What Is Resilience and How Can It Be Nurtured? A Systematic Review of Empirical Literature on Organizational Resilience. *International Journal of Health Policy and Management* 7 (6):491–503. doi: 10.15171/ijhpm.2018.06

Bathelt, H., and S. Henn. 2014. The Geographies of Knowledge Transfers over Distance: Toward a Typology. *Environment and Planning A: Economy and Space* 46 (6):1403–24. doi: 10.1068/a46115

Beckert, J. 1996. What is sociological about economic sociology? Uncertainty and the embeddedness of economic action. *Theory and Society* 25:803–40. doi: 10.1007/BF00159817

Blažek, J. 2016. Towards a typology of repositioning strategies of GVC/GPN suppliers: the case of functional upgrading and downgrading. *Journal of Economic Geography* 16 (4):849–69. doi: 10.1093/jeg/lbv027

Brill, F. N., and E. Robin. 2020. The risky business of real estate developers: network building and risk mitigation in London and Johannesburg. *Urban Geography* 41 (1):36–54. doi: 10.1080/02723638.2019.1637211

Brinks, V., and O. Ibert. 2020. From Corona Virus to Corona Crisis: The Value of An Analytical and Geographical Understanding of Crisis. *Tijdschrift voor economische en sociale geografie* 111 (3):275–87. doi: 10.1111/tesg.12428

Bristow, G., and A. Healy. 2014. Regional Resilience: An Agency Perspective. *Regional Studies* 48 (5):923–35. doi: 10.1080/00343404.2013.854879

Bryson, J. R. 2024. Jenga Capitalism and the Cyber-Energy-Production Plexus: New forms of Risk and Uncertainty that are Reshaping Economic Geography. In *The Changing Economic Geography of Companies and Regions in Times of Risk, Uncertainty, and Crisis*, ed. T. Neise, P. Verfürth, and M. Franz. Routledge: London.

Bryson J. R., and V. Vanchan. 2020. Covid-19 and Alternative Conceptualisations of Value and Risk in GPN Research. *Tijdschrift voor economische en sociale geografie* 111 (3):530–42. doi: 10.1111/tesg.12425

Buchheim, L., J. Dovern, C. Krolage, and S. Link. 2020. Firm-Level Expectations and Behavior in Response to the COVID-19 Crisis. CESifo Working Paper No. 8304. https://ssrn.com/abstract=3603773

Buck, C., T. Kreuzer, A. M. Oberländer, M. Röglinger, and M. Rosemann. 2022. Four Patterns of Digital Innovation in Times of Crisis. *Communications of the Association for Information Systems 50*:617–45. doi: 10.17705/1CAIS.05029

Dahles, H., and T. P. Susilowati. 2015. Business Resilience in Times of Growth and Crisis. *Annals of Tourism Research* 51:34–50. doi: 10.1016/j.annals.2015.01.002

David, L. 2018. Agency and Resilience in the Time of Regional Economic Crisis. *European Planning Studies* 26 (5):1041–59. doi: 10.1080/09654313.2018.1448754

Ebersberger, B., and A. Kuckertz 2021. Hop to it! The impact of organization type on innovation response time to the COVID-19 crisis. *Journal of Business Research 124*:126–35. doi: 10.1016/j.jbusres.2020.11.051

Ekinsmyth, C. 2002. Project Organization, Embeddedness and Risk in Magazine Publishing. *Regional Studies* 36 (3):229–43. doi: 10.1080/00343400220122043

Franz, M., T. Neise, and P. Verfürth. 2023. Gastronomie zwischen Strukturwandel und COVID-19-Krise: Handlungsmöglichkeiten für Wirtschaftsförderung und Stadtmarketing. In *Wirtschaftsförderung in der Krise: Konzepte zur Krisenbewältigung und Chancennutzung*, ed. T. Korn, J. Lempp, and G. van der Beek, 39–57. Wiesbaden: Springer Fachmedien Wiesbaden. doi: 10.1007/978-3-658-41390-3_4

Franz, M., T. Neise, and P. Verfürth. 2024. Learning from the COVID-19 crisis - how can business development and city marketing agencies support the restaurant and bar industry? In *The Changing Economic Geography of Companies and Regions in Times of Risk, Uncertainty, and Crisis*, ed. T. Neise, P. Verfürth, and M. Franz. Routledge: London.

Gong, H., R. Hassink, J. Tan, and D. Huang. 2020. Regional Resilience in Times of a Pandemic Crisis: The Case of COVID-19 in China. *Tijdschrift voor economische en sociale geografie* 111 (3):497–512. doi: 10.1111/tesg.12447

Haraguchi, M., T. Neise, W. She, and M. Taniguchi. 2023. Conversion strategy builds supply chain resilience during the COVID-19 pandemic: A typology and research directions. *Progress in Disaster Science* 17:100276.

Henig D., and D. M. Knight. 2023. Polycrisis: prompts for an emerging worldview. *Anthropology Today* 39(2):3–6. doi: 10.1111/1467-8322.12793

Jicha, F., Hassink, R., and H. Chu. 2024. Advancing regional economic resilience theorizing by integrating global value chain resilience. In *The Changing Economic Geography of Companies and Regions in Times of Risk, Uncertainty, and Crisis*, ed. T. Neise, P. Verfürth, and M. Franz. Routledge: London

Kalkuhl, M., and L. Wenz. 2020. The impact of climate conditions on economic production. Evidence from a global panel of regions. *Journal of Environmental Economics and Management* 103:102360. doi: 10.1016/j.jeem.2020.102360

Kitsos, A., and P. Bishop. 2018. Economic Resilience in Great Britain. The Crisis Impact and Its Determining Factors for Local Authority Districts. *Annals of Regional Science* 60 (2):329–47. doi: 10.1007/s00168-016-0797

Knight, F. H. 1921. *Risk, Uncertainty, and Profit*. Boston: Houghton Mifflin.

Lawrence, M., S. Janzwood, and T. Homer-Dixon. 2022. 'What Is a Global Polycrisis?' Version 2.0. Discussion Paper 2022-4. Cascade Institute. Accessed February 7, 2024. https://cascadeinstitute.org/technical-paper/what-is-a-global-polycrisis/

Liadze, I., C. Macchiarelli, P. Mortimer-Lee, and P. Sanchez Juanino. 2023. Economic costs of the Russia-Ukraine war. *The World Economy* 46(4):874–86. doi: 10.1111/twec.13336

Lüder, L., and L. Kalvelage. 2023. Regional resilience and social-ecological systems: the impact of COVID-19 on community conservation in Namibia. *African Geographical Review*. Advance online publication. doi: 10.1080/19376812.2023.2276939

Martin, R., and P. Sunley. 2020. Regional economic resilience: evolution and evaluation. In *Handbook on Regional Economic Resilience*, ed. G. Bristow and A. Healy, 1st ed., 10–35. Cheltenham: Edward Elgar.

Męczyński, M., and P. Ciesiółka. 2024. Towards a green transformation based on eco-innovations of enterprises in old industrial areas: The case of Silesia Province. In *The Changing Economic Geography of Companies and Regions in Times of Risk, Uncertainty, and Crisis*, ed. T. Neise, P. Verfürth, and M. Franz. Routledge: London.

Mol, M. J., L. Rabbiosi, and G. D. Santangelo. 2023. Should I stay or should I go?: how Danish MNEs in Russia respond to a geopolitical shift. *AIB Insights* 23(1):1–8. doi: 10.46697/001c.68337

Neise, T. 2024. Risk Narratives and Reconfiguration of Global Production Networks – The Case of the German Medical Technology Industry. In *The Changing Economic Geography of Companies and Regions in Times of Risk, Uncertainty, and Crisis*, ed. T. Neise, P. Verfürth, and M. Franz. Routledge: London.

Neise, T., T. López, and A. F. Angga Reksa. 2023. Rethinking Labour Risk in Global Production Networks: Resilience Strategies of Cruise Ship Workers in the Wake of the Covid-19 Pandemic. *Geoforum* 145:103842. doi: 10.1016/j.geoforum.2023.103842

Neise, T., M. T. Sambodo, and J. Revilla Diez. 2021. Are micro-, small-and medium-sized enterprises willing to contribute to collective flood risk reduction? Scenario-based field experiments from Jakarta and Semarang, Indonesia. *Organization & Environment* 34 (2), 219–42. doi: 10.1177/1086026619875435

November, V. 2008. Spatiality of Risk. *Environment and Planning A: Economy and Space* 40 (7):1523–27. doi: 10.1068/a4194

Onyeaka, H., C. K. Anumudu, Z. T. Al-Sharify, E. Egele-Godswill, and P. Mbaegbu. 2021. COVID-19 pandemic: A review of the global lockdown and its far-reaching effects. *Science Progress* 104(2):00368504211019854. doi: 10.1177/00368504211019854

Pachura, P. 2024. Risk perception of foreign businesses in the face of local contextual conditions. In *The Changing Economic Geography of Companies and Regions in Times of Risk, Uncertainty, and Crisis*, ed. T. Neise, P. Verfürth, and M. Franz. Routledge: London.

Rodrigue, J.-P., T. Notteboom, and A. A. Pallis. 2011. The financialization of the port and terminal industry: revisiting risk and embeddedness. *Maritime Policy & Management* 38 (2):191–213. doi: 10.1080/03088839.2011.556675

Rosenthal, U., M. T. Charles, and P. 't Hart. 1989. *Coping with crises: The management of disasters, riots and terrorism*. Springfield: Charles C. Thomas Publishers.

Schwabe, J. 2020. Risk and counter-strategies: The impact of electric mobility on German automotive suppliers. *Geoforum* 110:157–67. doi: 10.1016/j.geoforum.2020.02.011

Skjærseth, J. B., G. Bang, and M. A. Schreurs. 2013. Explaining growing climate policy differences between the European Union and the United States. *Global Environmental Politics* 13 (4):61–80. doi: 10.1162/GLEP_a_00198

Soroka, A., G. Bristow, M. Naim, and L. Purvis. 2020. Measuring regional business resilience, *Regional Studies* 54 (6):838–50. doi: 10.1080/00343404.2019.1652893

Terhorst, P., and H. Erkuş. 2024. Regional Economic Resilience, Uncertainty, and Fictional Expectations. In *The Changing Economic Geography of Companies and Regions in Times of Risk, Uncertainty, and Crisis*, ed. T. Neise, P. Verfürth, and M. Franz. Routledge: London.

Tooze, A. 2022. Welcome to the World of the Polycrisis. Accessed February 11, 2024. https://www.ft.com/content/498398e7-11b1-494b-9cd3-6d669dc3de33

Trippl, M., S. Fastenrath, and A. Isaksen. 2023. Rethinking regional economic resilience: Preconditions and processes shaping transformative resilience. *European Urban and Regional Studies*. Advance online publication. doi: 10.1177/09697764231172326

Van Egdom, D. C. Spitzmueller, X. Wen, M. A. Kazmi, E. Baranski, R. Flin, and R. Krishnamoorti. 2022. Job insecurity during an economic crisis: the psychological consequences of widespread corporate cost-cutting announcements. *Occupational health science* 6:1–25. doi: 10.1007/s41542-021-00102-8

Verfürth, P., T. Neise, and M. Franz. 2024. Contemporary perspectives on uncertainties, risks, and crises in Economic Geography. In *The Changing Economic Geography of Companies and Regions in Times of Risk, Uncertainty, and Crisis*, ed. T. Neise, P. Verfürth, and M. Franz. Routledge: London.

Völlers, P., T. Neise, P. Verfürth, M. Franz, F. Bücken, and K. P. Schumacher. 2023. Revisiting Risk in The Global Production Network Approach 2.0 - Towards A Performative Risk Narrative Perspective. *Environment and Planning A: Economy and Space* 55 (8):1838–58. doi: 10.1177/0308518X231169288

Völlers, P., N. Yavan, and M. Franz. 2021. The reaction of foreign firms to institutional changes: The case of German direct investment in Turkey. *Applied Geography* 134:102503. doi: 10.1016/j.apgeog.2021.102503

Yavan, N., M. R. Kalelioğlu, C. Kurtar Anlı. 2024. Risks and Opportunities in Digital Transformation of Industry in Peripheral Regions: The Case of Gaziantep, Turkey. In *The Changing Economic Geography of Companies and Regions in Times of Risk, Uncertainty, and Crisis*, ed. T. Neise, P. Verfürth, and M. Franz. Routledge: London.

Section A

Conceptual Discussions of Uncertainties, Risks, and Crises within Economic Geography

2 Contemporary Perspectives on Uncertainties, Risks, and Crises in Economic Geography

Philip Verfürth, Thomas Neise, Philip Völlers, Felix Bücken, and Martin Franz

2.1 Introduction

Economic geography is a dynamic research field that is grappling with the major challenges of our time, such as environmental changes (e.g., Neise, Sambodo, and Revilla Diez 2021), socio-political shifts (e.g., Völlers, Yavan, and Franz 2021), and shocking events (e.g., Brinks and Ibert 2020). In these largely interdisciplinary research areas, uncertainties, risks, and crises have become key terms, reflecting the constantly changing economic systems and their spatial patterns.

The terms uncertainties, risks, and crises are closely related. Uncertainties and risks are inherent features of economic activities and can result from various situations such as technological change, political instability, and natural disasters. These uncertainties and risks may manifest themselves in crises in which economic systems are disrupted, often leading to severe consequences for individuals and communities. The relationship between uncertainties, risks, and crises is thus a sequential one in which uncertainties and risks contribute to generating crises. Managing uncertainties and risks, for instance, through resilience strategies, is thus an essential factor in dealing with crises (Yeung 2023).

This chapter explores these fundamental terms to shed light on their conceptual foundations and examine their significance in the context of economic geography literature. Drawing on interdisciplinary insights from economics, sociology, organizational studies, and political science, the conceptual nuances of these terms and their theoretical foundations are outlined. By means of a review of seminal and contemporary literature, we elucidate how these concepts have been applied in the research fields of economic geography.

It is noticeable that there are clear differences in the scope and conceptual depth of the literature between risks, crises, and uncertainties. While there is extensive empirical work on risks, in particular, the concept of uncertainties has been far less intensively discussed and utilized. Accordingly, the discussion of the three concepts of uncertainties, risk, and crises will also vary in the following.

DOI: 10.4324/9781003457626-3

2.2 Conceptualizing Uncertainty

Uncertainty is a topic that has been discussed for a long time in a variety of disciplines. However, there is still an ongoing debate on the definition and understanding of uncertainty. One of the most important conceptual contributions to dealing with uncertainty in an economic context has been the distinction between risk and uncertainty by Knight (1921).

Knight (1921) distinguishes between risk and uncertainty by the possibility of assigning probabilities to certain situations in the economy. Risks, in his view, refer to conditions to which probabilities can be assigned. They can be calculated ex ante or based on statistics from the past and can be used as a basis for strategic decisions and managerial actions. Uncertainty, on the other hand, refers to conditions where individuals cannot rely on probability calculations. Situations of uncertainty are often unique and, therefore, associated with unknown outcomes (Beckert 1996), meaning that "we cannot predict or foresee what will happen when acting or not acting" (Aspers 2018, 133).

Even if Knight's distinction between uncertainty and risk is analytically clear-cut, it is difficult for actors to determine whether certain situations are uncertain or risky. It can be assumed that even situations that seem predictable due to probability distributions contain elements of uncertainty. This illustrates the limits of rational calculation of economic situations, and the importance of uncertainty in the modern economy (Beckert 2013; Mousavi and Gigerenzer 2014). Two main types of uncertainties can be recognized from the literature, which have different implications for the economic actions of decision-makers.

First, *informational uncertainty* arises due to a lack of available information or incomplete knowledge about a subject, event, or situation needed for a specific decision. An example of this would be when managers do not have complete information on location factors before deciding to invest in a new subsidiary. This type of uncertainty arises because of incomplete data or information as well as a lack of transparency (Vignoli et al. 2020).

Second, *fundamental uncertainty* is not solely due to a lack of information; it is rooted in the nature of the underlying processes or events with their unknown future outcomes. Fundamental uncertainty is based above all on the unpredictability of human behavior, and the fact that one's own economic actions always depend on third parties (Bathelt and Glückler 2003). It implies that in fundamentally uncertain situations relevant information "does not exist at the decision time because the future is yet to be created" (Dequech 2000, 41). An example of this would be assessing the economic consequences of political events such as geopolitical changes, where the outcome is unknown.

While economic theory traditionally viewed uncertainty as a threat to economic processes, alternative views have become established in organizational theory. In this view, uncertainty does not necessarily have to be reduced but can be used productively, for example, in engaging with creative processes (Ibert et al. 2021).

2.3 Uncertainty in Economic Geography Literature

Uncertainty has led to considerable debate in many areas of the social sciences. However, this is only partly the case in human geography (Senanayake and King 2021) and is far more evident in economic geography research. The limited studies that deal directly with uncertainty go across the sub-disciplines of economic geography (e.g., Glückler 2005; Ibert et al. 2021; Sohns and Wójcik 2020). They are mainly associated with a focus on situations and processes of major changes, and crises, as well as innovation and creativity, which are primarily characterized by fundamental uncertainty. Selected studies are presented below as illustrative examples.

How firms deal with uncertainty in processes of major changes is investigated by Glückler (2005), using the case of internationalization in the management consulting industry. The internationalization of management consulting firms is characterized by a high degree of uncertainty, fueled by three sources: the liabilities of foreignness, a lack of institutional standards, and the nature of consulting services as experienced goods. To successfully embed themselves in foreign local markets consulting firms must find mechanisms to bridge uncertainty. Two mechanisms were highlighted by Glückler (2005): trust and reputation. Personal, and especially goodwill trust, has been shown to be an important mechanism in building relational embeddedness, and to overcome uncertainty. Reputation, especially networked reputation, "has proved to be a key mechanism of social communication that reduces uncertainty, creates new relations between formerly unrelated actors, and thus facilitates the emergence of structural embeddedness" (Glückler 2005, 1746). Overall, the study shows the importance of social practice institutions in the uncertain internationalization processes of firms.

How firms mediate uncertainty generated by crises has recently drawn attention in economic geography research, especially due to the case of Brexit (e.g., Fuller 2023; Panitz and Glückler 2022; Sohns and Wójcik 2020). Fuller (2023) analyzed this question using the case of foreign subsidiaries in the Southeast of England and Wales. He shows that the resulting responses of foreign subsidiaries caused by Brexit are geographically uneven. The research demonstrates that foreign subsidiaries in Southeast England can cope with the high degree of uncertainty induced by Brexit. Fuller attributes this to the companies' status as high-value creation subsidiaries, which goes hand in hand with autonomy and the ability to take independent measures. In contrast, foreign subsidiaries in Wales have less autonomy and ability to react independently to the crisis, due to their status as low-value creation subsidiaries. The study emphasizes the important role of fictional expectations (Beckert 2013) in mediating uncertainty in crises. It is these cognitive or narrative frames that provide decision-makers guidance for action to navigate the unknown outcomes of the future.

Innovation and creativity, as time-spatial processes with unknown outcomes, are also part of the economic-geographical debate on uncertainty.

Ibert et al. (2021) analyze the role of uncertainty in creative collaborations in the fields of music production and pharmaceutical development. The peculiarity of this study is that the authors do not conceptualize uncertainty as a threat to economic interaction but as a productive asset. Using the example of eight creativity biographies, three central practices in dealing with uncertainty are identified: embracing, fixing, and ignoring.

Embracing uncertainty can be understood as a practice of actors adopting an attitude of open-mindedness as well as receptiveness to surprises and unpredictabilities. Actors deal actively with uncertainty in creative processes by consciously venturing into unexplored areas and value open-ended exploration. Therefore embracing "encompasses awareness and acceptance of possible failure, and a general tendency to observing and reacting, rather than planning" (Ibert et al. 2021, 124).

Fixing uncertainty involves dealing with situations where multiple options of similar value and viability exist, often occurring in ambiguous scenarios. Participants assign a temporary resolution to one uncertain aspect, allowing the focus to shift to another uncertain dimension. These fixes guide decision-making temporarily, but they influence subsequent steps and may halt exploration in certain directions.

By ignoring aspects of uncertainties, participants of creative processes act "rather with a general disposition of confidence toward established or upcoming uncertainty, in the sense of 'pretending' or acting as if an already identified uncertainty does not exist" (Ibert et al. 2021, 128). In other words, actors prefer to be ignorant, especially when no plausible solutions are available.

2.4 Conceptualizing Risks

The term risk is characterized by its multitude of definitions and numerous changes in meaning. Thus, it is essential to highlight its conceptual particularities. Above all, it is important to emphasize that risks—unlike dangers—are always the result of a decision made beforehand. In the vein of a social constructivist view on risk, it is the decision that leads to the possibility of creating an undesirable state, thus, risks can be understood as the product of human action (Zehetmair 2012). Accordingly, decision-making is of cardinal importance for risk research, as the impression that the world is becoming increasingly risky for the people who live in it (Beck 2008) is primarily due to the increasing complexity of the consequences of decisions. The globalization of supply chains, insecure critical infrastructures, and risks arising from cultural differences, make it difficult to assess the consequences of economic decisions (Kwak, Seo, and Mason 2018; Pachura 2024 in this book). In addition, many risks are highly interdependent. For example, legal frameworks, environmental impact, market, and capital access, reputations, and resistance, influence each other in complex ways (e.g., Franz, Schlitz, and Schumacher 2018; Starr, Newfrock, and Delurey 2003). Risks can also undergo dynamic shifts; depending on environmental changes, previously latent risks can become acute. At the same time, risks that were very prominent over a certain period can recede into the background and regress to a latent stage (Barnett and Pratt 2000).

At first glance, the concept of risk, in line with Knight (1921), is inherently associated with the desire for objectivity in predicting future events using quantitative methods: "Risk prevails in the 'world of classical statistics,' where there is a known set of possible future events and every event can be attributed a probability (distribution) based on observations about relative frequencies with which these events took place in the past" (Bartkowski and Hansjürgens 2019, 21). On this basis, the so-called risk formula (product of probability and consequence) is used to objectify risks. The formula enables the assessment of a situation at a specific point in time, independent of its context and comparison with other configurations (Renn 1990). Furthermore, it not only allows for quantitative comparisons of different risks, but also the calculation of the effects of protective measures (Zehetmair 2012).

However, there are methodological limitations to quantifying the effects of damages, faults, and impairments which need to be accepted. Not all impacts of harmful events can be described through feature counts, nor can they be expressed in financial terms. Thus, classic variables used in the analysis of environmental disasters (e.g., number of victims, damage in dollars) do not provide a complete understanding of the consequences of events. Effects such as disrupted family relationships, post-traumatic stress disorders, or depression among disaster victims, are hardly able to be grasped through quantitative approaches (Rohrmann 2008). The same applies to the individual perception of risks. As the mere fact that a risk is classified as highly relevant based on statistical indicators, does not mean that it also guides action; it is worthwhile to give increased attention to this aspect (Renn 1990). This is necessary not least because empirical comparisons between quantitative risk assessments and individual risk perceptions often reveal significant discrepancies (Greiving et al. 2021). The basis for differing perceptions lies in various interpretations of the world, which, in turn, are shaped by experiences and strong beliefs that arise within the context of norms, values, and cultural characteristics (Lupton 2013; Slovic 1992). At the same time, social processes in the actors' environment (such as the influence of the mass media) serve to amplify certain risks while seemingly neglecting others (Pidgeon and Beattie 1998).

For a deeper understanding of economic activity in a spatial context, it is important to note that the risk perception of individuals is also relevant at the organizational level. Depending on their position or competencies within the organization, the personal risk perception of managers directly influences the organizational response to risks (Völlers, Yavan, and Franz 2021). This explains why companies can have very different strategies for dealing with comparable risks and in similar situations (Franz, Schlitz, and Schumacher 2018).

2.5 Risk in Economic Geography Literature

Debates on risk are mainly situated in three sub-disciplines of economic geography, within the domain of financial geography (e.g., Ashton 2009; Clark 2010;

Johnson 2013), in research on globalization processes (e.g., Bathelt and Henn 2021; Coe and Yeung 2015; Torrance 2009), and environmental economic geography (e.g., Neise, Revilla Diez, and Garschagen 2018; Pollard et al. 2008; Stanley 2013). Most prominently debated are two types of risk: financial risk (e.g., Bennett 2009; Christophers, Bigger, and Johnson 2020; Clark 2010) and risk related to the global interlinking of economic activities (e.g., Bryson and Vanchan 2020; Coe and Yeung 2015; Völlers et al. 2023).

A number of economic geographers, directly or indirectly, refer to the economic notion of risk defined by Knight (1921), who differentiates between risk (calculable) and uncertainty (non-calculable). Although a critical engagement with the risk concept is becoming increasingly apparent, the objectifying notion of risk is still predominantly applied in subsequent debates—following a risk-and-reward-paradigm, or a trade-off-thinking between risk and opportunities, in favor of opposing terms such as risk and reward (e.g., Blažek 2016), risk and value (e.g., Bryson and Vanchan 2020), as well as risk and trust (e.g., Bathelt and Henn 2014). However, debates in economic geography literature are also influenced by sociological and cultural political economy perspectives on risk, acknowledging the role of microfoundational aspects.

Some authors start emphasizing an actor-centered stance with socio-spatial emphasis. In so doing, these economic-geographical scholars stress variables such as spatial-relation context (e.g., Ekinsmyth 2002; November 2008; Rodrigue, Notteboom, and Pallis 2011) and more specifically embeddedness, and the role of risk perception (e.g., Gagliardi and Iammarino 2018; Kleibert 2015; Völlers 2023), in addition to connecting the conceptualization of risk to the idea of future-oriented expectations (e.g., Baldwin and Stanley 2013; Christiansen 2021; Green 2000). These debates build alternative conceptual arguments regarding the nature of risk, which is seen as resulting from social practices rather than as an objective entity (e.g., Brill and Robin 2020; Schwabe 2020; Völlers et al. 2023). Especially studies emphasizing the fictional aspect (see Beckert 2013) of humans fabricating *the* future, recognizing facets of risk production such as power relations, legitimacy, governmentality, and the discursive or narrative (e.g., Emel and Huber 2008; Green 2000; Lanari et al. 2021), and social-cultural aspects of reputation-related risks (e.g., Clark and Hebb 2005; Ibert et al. 2019; Kleibert 2021). Additionally, some authors engage in debates on risk while drawing attention to the mediating role of formal (regulations) and informal (interpersonal networks) institutions (Bathelt and Henn 2014; Bennett 2000; Rafiqui 2009).

Economic-geographical literature focusing on individual human behavior and decision-making, specifically highlights the role of context and cognition (e.g., Clark 2022; Strauss 2008, 2009). One of the early accounts adopting a constructivist perspective on risk in economic-geographical literature represents the case study of the environmental understanding and knowledge of local (lay) people regarding industrial hazards in England by Irwin, Simmons, and Walker (1999). The authors advocate for acknowledging the context-specific sense-making processes of how individuals interpret their immediate surroundings in everyday life. "[...] [R]isk does not stand apart from the range of social relationships, worldviews, everyday practices, and shared understandings which constitute local culture" (Irwin, Simmons, and Walker 1999, 1325). In the same vein, November (2008, 1526) argues for a spatial aspect in the analysis of risk perception and assessments as "space influences the implications of risk just as risks affect and alter spaces themselves" (see also Stanley 2013).

An increasing number of economic geographers are analyzing risk-related globalization processes through the lens of the so-called Global Production Networks approach 2.0 (GPN 2.0), with an underlying network perspective on cross-border economic activities. This GPN 2.0 discussion started to gain momentum when Coe and Yeung (2015), introduced the concept of GPN risk. In this way, Coe and Yeung (2015, 110) argue that "global production networks are fundamentally an organizational platform for economic actors to mitigate the different forms of *risk* inherent in the competitive dynamics of cost reduction, capability building, market growth, and financial discipline." The authors differentiate between five types of risk: economic, product, regulatory, labor, and environmental. Thereby, four characteristics were identified by Coe and Yeung (2015) describing GPN risk: First, risk has the potential to be transmissive and amplifiable through coupled locations and interlinked GPN actors. Second, the nature of coordinative management approaches is determined by the fundamental nature of risk. Third, applied risk management grids allow deductions to be made about the power relations within a network. Fourth, certain risks are specific to particular industry sectors (e.g., Coe 2021; Yeung 2022). In principle, GPN risk is conceptualized as being produced externally by GPN actors. However, other authors argue for including different scales to consider where risk is produced (e.g., Bryson and Vanchan 2020; Völlers et al. 2023; Völlers 2023). Geenen (2018) and Völlers et al. (2023) highlight the socially constructive nature of risk and its effects on network dynamics and organizational structures—"risk as being politically produced" (Geenen 2018, 32). Barratt and Ellem (2019) stress the spatio-temporal component of assessing and managing network-related risk dynamics; additionally, the authors draw attention to the fact that a chosen risk management strategy may itself become a risk at a later point in time.

2.6 Conceptualizing Crisis

For a long time, crises were seen as an external element beyond the control of people, organizations, and states. A crisis was seen as an "act of God that is unwanted, unexpected, unprecedented, and almost unmanageable, causing widespread unbelief and uncertainty" (Rosenthal, Boin, and Comfort 2001, 5). Since the beginning of the 21st century, scholars have been increasingly concerned with the emergence, prevention, and management of crises. The understanding of crisis in today's literature is that it is not an external feature, but has been internalized in individual life and economic, social, and ecological systems (Rosenthal et al. 2001).

Nevertheless, the term crisis remains a multifaceted concept. However, a common understanding is that crises are "a serious threat to the basic structures or the fundamental values and norms of a system, which under time pressure and highly uncertain circumstances necessitates making critical decisions" (Rosenthal, Charles, and 't Hart 1989, 10). Roitman (2021, n. p.) stresses that crises "create narrative structure: it is a means to define an event and to delineate a beginning and an end." Calling crises a normative concept, she emphasizes that scholars should focus on the complexity of crisis discourses and understand how crisis debates are socially constructed and by whom (Roitman 2014).

Furthermore, there is a general consensus that crises are not purely discrete events, but rather unfolding as a process with certain stages (Rosenthal et al. 2001). Several scholars and organizations (e.g., Boin et al. 2005; Kruke 2012, UNISDR 2009) categorize crises into different phases, including mitigation, preparedness, prevention, response, and recovery. These "phase" models have been questioned, as they tend to oversimplify the complexity of crisis (Coetzee and van Niekerk 2012), thus leading to the conceptual development of circular models. Kruke (2012), for instance, suggests that a new normal emerges during a pre-crisis (including preparedness and prevention), and undergoes an acute (including response) and a post-crisis phase (including recovery and learning) (Morsut et al. 2021).

Next to stressing the stages covering a moment of crisis, research has engaged in classifying various types of crises (Drennan et al. 2015; McConnell 2003; t'Hart and Boin 2001). McConnell (2003) describes three different forms: sudden, chronic, and creeping crises. Sudden crises are characterized by their unexpected and rapidly evolving nature. In contrast, chronic crises are persistent and often enduring, lacking clear-cut solutions. Creeping crises gradually escalate over an extended period, spanning years or even decades. t'Hart and Boin (2001) offer a comparable categorization, which, however, focuses primarily on the speed of onset and completion. The authors distinguish between fast-burning, cathartic, slow-burning, or long-shadow crises. Like the term sudden crisis, fast-burning ones entail swift and conclusive beginnings and endings, sometimes with tragic outcomes. The cathartic crisis, resembling McConnell's "creeping crisis," evolves gradually, but concludes abruptly. A slow-burning crisis, corresponding to McConnell's chronic crisis, unfolds gradually in both onset and progression, often dissipating rather than ending definitively. A long-shadow crisis implies that the effects and repercussions of this crisis linger over an extended period, casting a figurative shadow that persists over time. These crises may unfold gradually, often building up over time before reaching a critical point, and their impacts can be enduring.

Although crises result in harmful consequences, there is also the understanding that "crises are about change [...] crises are opportunities to impose new ideas and practices" (French and Leyshon 2010, 2549). Particularly, in economic-orientated science, crises are seen—next to their negative effects—as a critical component for renewal and innovation. In line with Schumpeter's idea of "creative destruction," a crisis can be a critical juncture to rethink or break inefficient development paths (McCraw 2009).

As crises are increasingly spreading geographically, or several harmful events take place simultaneously or interact with each other in the era of globalization, several new conceptualizations have been developed. Boin (2009), for instance, emphasizes the term transboundary crisis, which transcends conventional time and geographical boundaries. Moreover, their origins can be profound, and their effects may only become apparent years later (Boin 2009). More recently, literature has moved its focus to the term poly-crises. Lawrence, Janzwood, and Homer-Dixon (2022), for instance, call the occurrence of crises in multiple global systems that are causally linked, a poly-crisis. Due to their interconnectedness, crises have a synergistic effect that causes greater damage than the sum of the individual crises on their own (see also Henig and Knight 2023; Tooze 2022).

2.7 Crises in Economic Geography Literature

The term crisis has been approached from different angles in economic geography, highlighting various key events (e.g., global financial crisis in 2007/08, European austerity policies, Brexit, and the COVID-19 pandemic) with profound effects on spatial relations and uneven development.

Related to structural and economic crises, the term resilience has received increasing attention from economic geography scholars. In particular, the global financial crisis of 2007/08 and its aftermath, have led to the concept of resilience being incorporated into debates in economic geography. Resilience, defined as the ability of a system to recover from shocks, has been widely adopted in understanding how regions, sectors, or economic systems resist and recover from various crises (Gong et al. 2020; Hassink 2010; Martin 2018; Simmie and Martin 2010). Many studies have taken a quantitative approach to analyze how crises lead to different regional growth paths. This approach relates to an understanding of resilience as a bouncing back effect, in other words returning to pre-shock conditions (e.g., Christopherson, Michie, and Tyler 2010; Dijkstra, Garcilazo, and McCann 2015), or a bounce-forward approach that focuses on reorientation or developing new growth paths (e.g., Boschma 2015; Martin et al. 2016).

Due to contemporary instances of crisis, such as the COVID-19 pandemic, the resilience concept has been included more deeply in economic geography, resulting in a larger diversity of conceptual approaches. For instance, scholars have recently developed the concept of *transformative resilience* that concentrates on a region's capacities to transform their paths toward sustainability (Martin and Sunley 2020; Trippl, Fastenrath, and Isaksen 2023). In addition, several scholars have explored the resilience of regions and sectors during crises through concepts such as *regional resilience* (e.g., Gong et al. 2020; Lüder and Kalvelage 2023), *business resilience* (e.g., Soroka et al. 2020) or *organizational resilience* (e.g., Verfürth et al. 2022). The importance of local characteristics in building resilience during or after crises has been highlighted, indicating that resilience is not a one-size-fits-all concept, and is influenced by specific contextual factors and actors' agency (Bristow and Healy 2014; David 2018; Kitsos and Bishop 2018).

The critical events in the 21st century also influenced financial geography, a subdiscipline of economic geography (e.g., Engelen and Faulconbridge 2009; Klagge 2009; Lee et al. 2009). Research has drawn attention to the spatial architecture and power relations of the global finance industry to understand the origins, unfolding, and aftermath of crises, as well as their impact on spatial inequalities (e.g., Christophers 2015; Martin 2011; Muellerleile et al., 2014; Omstedt 2016; Wójcik 2013). Furthermore, a large body of research has focused on the varying impacts of financial crises on regions, the spatial dynamics of financial centers, and the role of financialization and mortgage crisis in exacerbating urban struggles (Dymski 2018; French, Leyshon, and Thrift 2009; Hall 2011; Martin 2011; Pažitka and Wójcik 2018).

With a specific focus on financial crises and the COVID-19 pandemic in particular, economic geography scholars have engaged in questioning the capitalist architecture of the global economy and labeled, for instance, the world financial crisis as a crisis of neoliberalism (e.g., Aalbers 2013; French, Leyshon, and Thrift 2009). In this context, the work of David Harvey (1985) on spatial fix, emphasizing the spatial dimensions of capitalism and crisis, is influential. His work highlights how spatial inequalities (e.g., resources, infrastructure) and uneven regional growth patterns can create vulnerabilities that amplify the impact of economic crises (Jones and Ward 2004; Oßenbrügge 2011). Harvey (1985, 11) sees crises as "the real manifestation of the underlying contradictions within the capitalist process of accumulation." Furthermore, Harvey (2003) distinguishes between partial, switching, and global crises. Partial crises only affect single sectors or geographical regions and may be solved within this sector or region. Global crises affect all sectors and regions of the capitalist world. More intriguing is the idea of geographical switching crises that occur as a result of the geographical reorganization of capitalism. According to Harvey, these crises are not caused by overaccumulation, as is traditionally understood, but rather are the effect of the spatial reconfiguration of capital within the capitalist system (Franz 2021; Harvey 2003).

As a result of the COVID-19 pandemic, the term crisis was associated with the dark side of economic geography. Scholars are looking at the unequal and exclusionary development outcomes of global production, exacerbated by crises such as the COVID-19 pandemic (Lawreniuk 2020; Neise, López, and Angga Reksa 2023; Sokol and Pataccini 2020). Likewise, the concept of polycrisis has started to be included in research concerning the fragility of global value chains or GPN (e.g., Gong et al. 2022).

Although moments of crisis have resulted in a large body of case studies in economic geography, there is less work on how the term crisis challenges economic geography concepts. An exception is the work by Hadjimichalis and

Hudson (2014, 210), who argue that the impact of the 2007/08 world financial crisis "pushed dominant regional development theories to an [sic] homologous deep theoretical crisis." The authors suggest a paradigm shift in regional development theories that moves away from a too narrow emphasis on markets, entrepreneurship, and competitiveness. They advocate for a more holistic view of how markets are politically constructed and socially regulated that includes socio-spatial justice, interventionist regional policies, as well as social and spatial redistribution. Hadjimichalis and Hudson (2014) argue that crisis is inherent to capitalism, and it is necessary to develop a systemic view of capitalism that acknowledges periodic crises as a necessary and recurrent feature of capitalist development. They also suggest understanding the spatial basis of crises and their impact on different regions, to better address the uneven development and spatial inequalities that crises exacerbate.

Another analytical focus on the term crisis within economic geography has been developed by Brinks and Ibert (2020). They emphasize the importance of understanding crisis as more than just a temporal event, and the characteristic elements of crises: uncertainty, urgency, and threat. Brinks and Ibert (2020) argue for geographic thinking in crisis studies by taking an actor-centric approach. Through the TPSN framework (territory, place, space, and network) developed by Jessop, Brenner, and Jones (2008), Brinks and Ibert (2020) provide a better understanding of how crises unfold in time and space, that helps to build effective crisis management and response measures with consideration of regional characteristics.

2.8 Conclusion

In this chapter, we have outlined contemporary perspectives of uncertainties, risks, and crises within the realm of economic geography (see Table 2.1). Through a conceptual lens, we explored the theoretical foundations and nuanced meanings of these fundamental terms. In addition, we elucidate how these concepts have been applied in empirical research within economic geography through a literature review.

Uncertainty is rooted in the nature of economic processes or events, with their unknown future outcomes. It can also stem from a lack of available information or incomplete knowledge about a subject, event, or situation needed for a specific decision. Uncertainty is conceptualized primarily from an objective perspective, as conditions for which no probabilities can be assigned (Knight 1921). Within economic geography, uncertainty has been examined across the sub-disciplines, but compared to risks and crises only to a limited extent. The studies dealing directly with uncertainty mainly focus on situations and processes of major changes, crises, innovation, and creativity.

The term risk is conceptualized either from an objective perspective, as conditions for which probabilities can be assigned, or from a social constructivist perspective (Völlers et al. 2023). The economic geography debate dealing with risk is largely taking place in the sub-disciplines of financial geography,

Table 2.1 Uncertainties, risks, and crises in Economic Geography.

Terms	*Conceptualization*	*Appearance in literature*
Uncertainties	Objective perspective on uncertainty: conditions for which no probabilities can be assigned. Types of uncertainty: e.g., informational and fundamental uncertainty.	A limited number of studies deal directly with uncertainty. Studies across the sub-disciplines of economic geography. Mainly associated with a focus on situations and processes of major changes, crises, innovation, and creativity.
Risks	Objective vs. social constructivist perspective on risk. Types of risk: e.g., financial, economic, product, regulatory, labor, and environmental risk.	A large variety of studies deal directly with risk. Studies mainly in three sub-disciplines: financial geography, geographies of globalization, and environment-oriented economic geography. Mainly associated with a focus on individual and organizational decision-making processes as well as on risk perception.
Crises	Objective vs. social constructivist perspective on crisis. Types of crises: e.g., fast-burning, cathartic, slow-burning, or long-shadow crises, and poly-crises.	A large variety of studies deal directly with crises. Studies across the sub-disciplines of economic geography, especially in the domain of financial geography. Mainly associated with a broad empirical interest, such as the origin of a crisis, resilience within a crisis, and impacts of a crisis.

Source: own elaborations.

geographies of globalization, and environmental economic geography. The main focus of these studies is on individual and organizational decision-making processes as well as on risk perception.

Crises are also conceptualized from an objective perspective, as a serious threat that requires critical decision-making (Rosenthal et al. 1989) and from a social constructivist perspective (Roitman 2021). Across the sub-disciplines of economic geography, there are scholars with a broad interest in studying crisis, focusing on the origin of crisis, resilience within crisis, as well as on the impacts of crisis.

Uncertainties, risks, and crises are, and will remain, important concepts in economic geography, linking diverse thematic issues such as processes of globalization, environmental changes, socio-political shifts, and technological advancements. We see a need for further research in two areas. First, while there is already an intensive theoretical and empirical debate on the concepts of risk and crises in economic geography, we encourage economic geographers to become more involved in conceptual debates about uncertainty and to conduct more empirical studies dealing directly with uncertainty from a spatial perspective. Second, future research could contribute to bridging the conceptual boundaries of uncertainties, risks, and crises, by focusing on the close connections and interrelationships of these terms and gaining new insights beyond the limitations of the singularly focused perspectives.

References

Aalbers, M. B. 2013. Debate on Neoliberalism in and after the Neoliberal Crisis. *International Journal of Urban and Regional Research* 37 (3):1053–57. doi: 10.1111/1468-2427.12061

Ashton, P. 2009. An Appetite for Yield: The Anatomy of the Subprime Mortgage Crisis. *Environment and Planning A: Economy and Space* 41 (6):1420–41. doi: 10.1068/a40328

Aspers, P. 2018. Forms of uncertainty reduction: Decision, valuation, and contest. *Theory and Society* 47 (2):133–49. doi: 10.1007/s11186-018-9311-0

Baldwin, A., and A. Stanley. 2013. Risky natures, natures of risk. *Geoforum* 45:2–4. doi: 10.1016/j.geoforum.2012.12.016

Barnett, C. K., and M. G. Pratt. 2000. From threat-rigidity to flexibility - Toward a learning model of autogenic crisis in organizations. *Journal of Organizational Change Management* 13:74–88. doi: 10.1108/09534810010310258

Barratt, T., and B. Ellem. 2019. Temporality and the evolution of GPNs: remaking BHP's Pilbara iron ore network. *Regional Studies* 53 (11):1555–64. doi: 10.1080/00343404.2019.1590542

Bartkowski, B., and B. Hansjürgens. 2019. Risk and Uncertainty as Sources of Economic Value of Biodiversity and Ecosystem Services. In *Atlas of ecosystem services: Drivers, risks, and societal responses*, ed. M. Schröter, S. Klotz, and R. Seppelt, 1st ed. 21–24. Basel: Springer.

Bathelt, H., and J. Glückler. 2003. Toward a relational economic geography. *Journal of Economic Geography* 3:117–44.

Bathelt, H., and S. Henn. 2014. The Geographies of Knowledge Transfers over Distance: Toward a Typology. *Environment and Planning A: Economy and Space* 46 (6):1403–24. doi: 10.1068/a46115

Bathelt, H., and S. Henn. 2021. Knowledge exchanges, trust, and secretive geographies in merger and acquisition processes. *Environment and Planning A: Economy and Space* 53 (6):1435–53. doi: 10.1177/0308518X211013360

Beck, U. 2008. *Weltrisikogesellschaft: Auf der Suche nach der verlorenen Sicherheit* [World risk society: In search of lost security]. Suhrkamp.

Beckert, J. 1996. What is sociological about economic sociology? Uncertainty and the embeddedness of economic action. *Theory and Society* 25:803–40.

Beckert, J. 2013. Imagined futures: Fictional expectations in the economy. *Theory and Society* 42 (3):219–40. doi: 10.1007/s11186-013-9191-2

Bennett, P. 2000. Mutuality at a Distance? Risk and Regulation in Marine Insurance Clubs. *Environment and Planning A: Economy and Space* 32 (1):147–63. doi: 10.1068/a3215

Bennett, P. 2009. Financial Risks and Management. In *International encyclopedia of human geography*, ed. R. Kitchin, and N. Thrift, 167–72. Amsterdam: Elsevier.

Blažek, J. 2016. Towards a typology of repositioning strategies of GVC/GPN suppliers: the case of functional upgrading and downgrading. *Journal of Economic Geography* 16 (4):849–69. doi: 10.1093/jeg/lbv027

Boin, A. 2009. The New World of Crises and Crisis Management: Implications for Policymaking and Research. *Review of Policy Research* 26 (4):367–77. doi: 10.1111/j.1541-1338.2009.00389.x

Boin, A., P. 't Hart, E. Stern, and B. Sundelius. 2005. *The Politics of Crisis Management: Public Leadership Under Pressure*. Cambridge: Cambridge University Press.

Boschma, R. 2015. Towards an Evolutionary Perspective on Regional Resilience. *Regional Studies* 49 (5):733–51. doi: 10.1080/00343404.2014.959481

Brill, F. N., and E. Robin. 2020. The risky business of real estate developers: network building and risk mitigation in London and Johannesburg. *Urban Geography* 41 (1):36–54. doi: 10.1080/02723638.2019.1637211

Brinks, V., and O. Ibert. 2020. From Corona Virus to Corona Crisis: The Value of An Analytical and Geographical Understanding of Crisis. *Tijdschrift voor economische en sociale geografie* 111 (3):275–87. doi: 10.1111/tesg.12428

Bristow, G., and A. Healy. 2014. Regional Resilience: An Agency Perspective. *Regional Studies* 48 (5):923–35. doi: 10.1080/00343404.2013.854879

Bryson J. R., and V. Vanchan. 2020. Covid-19 and Alternative Conceptualisations of Value and Risk in GPN Research. *Tijdschrift voor economische en sociale geografie* 111 (3):530–42. doi: 10.1111/tesg.12425

Christiansen, J. 2021. Fixing fictions through blended finance: The entrepreneurial ensemble and risk interpretation in the Blue Economy. *Geoforum* 120:93–102. doi: 10.1016/j.geoforum.2021.01.013

Christophers, B. 2015. Geographies of Finance II. Crisis, Space and Political-Economic Transformation. *Progress in Human Geography* 39 (2):205–13. doi: 10.1177/0309132513514343

Christophers, B., P. Bigger, and L. Johnson. 2020. Stretching scales? Risk and sociality in climate finance. *Environment and Planning A: Economy and Space* 52 (1):88–110. doi: 10.1177/0308518X18819004

Christopherson, S., J. E. Michie, and P. Tyler. 2010. Regional Resilience: theoretical and empirical perspectives. *Cambridge Journal of Regions, Economy and Society* 3 (2):3–10. doi: 10.1093/cjres/rsq004

Clark, G. L. 2010. Human nature, the environment, and behaviour: explaining the scope and geographical scale of financial decision-making. *Geografiska Annaler: Series B, Human Geography* 92 (2):159–73. doi: 10.1111/j.1468-0467.2010.00340.x

Clark, G. L. 2022. Agency, sentiment, and risk and uncertainty: Fears of job loss in 8 European countries. *ZFW – Advances in Economic Geography* 66 (1):3–17. doi: 10.1515/zfw-2021-0037

Clark, G. L., and T. Hebb. 2005. Why Should They Care? The Role of Institutional Investors in the Market for Corporate Global Responsibility. *Environment and Planning A: Economy and Space* 37 (11):2015–31. doi: 10.1068/a38116

Coe, N. M. 2021. *Advanced introduction to global production networks*. Cheltenham: Edward Elgar.

Coe, N. M., and H. W. Yeung. 2015. *Global production networks. Theorizing economic development in an interconnected world*. Oxford: Oxford University Press.

Coetzee, C., and D. van Niekerk. 2012. Tracking the evolution of the disaster management cycle: A general system theory approach. *Jàmbá: Journal of Disaster Risk Studies* 4 (1):1–9. doi: 10.4102/jamba.v4i1.54

David, L. 2018. Agency and Resilience in the Time of Regional Economic Crisis. *European Planning Studies* 26 (5):1041–59. doi: 10.1080/09654313.2018.1448754

Dequech, D. 2000. Fundamental uncertainty and ambiguity. *Eastern Economic Journal* 26 (1):41–60.

Dijkstra, L., E. Garcilazo, and P. McCann. 2015. The effects of the global financial crisis on European regions and cities. *Journal of Economic Geography* 15 (5):935–49. doi: 10.1093/jeg/lbv032

Drennan L. T., A. McConell, and S. Alastair. 2015. *Risk and Crisis Management in the Public Sector*. 2nd ed. London: Routledge.
Dymski, G. 2018. Finance and Financial Systems: Evolving Geographies of Crisis and Instability. In *The New Oxford Handbook of Economic Geography, Oxford Handbooks*, ed. G. L. Clark, M. Feldman, M. S. Gertler, and D. Wójcik, online edn, Oxford Academic, February 5, 2018. doi: 10.1093/oxfordhb/9780198755609.013.4
Ekinsmyth, C. 2002. Project Organization, Embeddedness and Risk in Magazine Publishing. *Regional Studies* 36 (3):229–43. doi: 10.1080/00343400220122043
Emel, J., and M. T. Huber. 2008. A risky business: Mining, rent and the neoliberalization of "risk". *Geoforum* 39 (3):1393–407. doi: 10.1016/j.geoforum.2008.01.010
Engelen, E., and J. Faulconbridge. 2009. Introduction: Financial Geographies-the Credit Crisis as an Opportunity to Catch Economic Geography's Next Boat? *Journal of Economic Geography* 9 (5):587–95. doi: 10.1093/jeg/lbp037
Franz, T. 2021. Spatial fixes and switching crises in the times of COVID-19: implications for commodity-producing economies in Latin America. *Canadian Journal of Development Studies* 42 (1/2):109–21. doi: 10.1080/02255189.2020.1832881
Franz, M., N. Schlitz, and K. P. Schumacher. 2018. Globalization and the Water-Energy-Food Nexus – Using the Global Production Networks Approach to analyze Society-Environment Relations. *Environmental Science & Policy* 90:201–12. doi: 10.1016/j.envsci.2017.12.004
French, S., and A. Leyshon. 2010. 'These f@#king guys': The terrible waste of a good crisis. *Environment and Planning A* 42 (11):2549–59. doi: 10.1068/a43421
French, S., A. Leyshon, and N. Thrift. 2009. A very geographical crisis: the making and breaking of the 2007–2008 financial crisis. *Cambridge Journal of Regions, Economy and Society* 2 (2):287–302. doi: 10.1093/cjres/rsp013
Fuller, C. 2023. Uncertainty, fictional expectations and economic agency. *Geoforum* 140:103699. doi: 10.1016/j.geoforum.2023.103699
Gagliardi, L. and S. Iammarino. 2018. Innovation in risky markets: ownership and location advantages in the UK regions. *Journal of Economic Geography* 18 (5):1177–201. doi: 10.1093/jeg/lby033
Gong H., R. Hassink, C. Foster, M. Hess, and H. Garretsen. 2022. Globalisation in reverse? Reconfiguring the geographies of value chains and production networks. *Cambridge Journal of Regions Economics and Society* 15:165–81. doi: 10.1093/cjres/rsac008
Gong, H., R. Hassink, J. Tan, and D. Huang. 2020. Regional Resilience in Times of a Pandemic Crisis: The Case of COVID-19 in China. *Tijdschrift voor economische en sociale geografie* 111 (3):497–512. doi: 10.1111/tesg.12447
Geenen, S. 2018. Underground dreams. Uncertainty, risk and anticipation in the gold production network. *Geoforum* 91:30–38. doi: 10.1016/j.geoforum.2018.02.019
Glückler, J. 2005. Making embeddedness work: social practice institutions in foreign consulting markets. *Environment and Planning A* 37 (10):1727–50. doi: 10.1068/a3727
Green, S. 2000. Negotiating with the Future: The Culture of Modern Risk in Global Financial Markets. *Environment and Planning D: Society and Space* 18 (1):77–89. doi: 10.1068/d205t
Greiving, S., M. Fleischhauer, C. D. León, L. Schödl, G. Wachinger, I. K. Quintana Miralles, and B. Prado Larraín. 2021. Participatory Assessment of Multi Risks in Urban Regions - The Case of Critical Infrastructures in Metropolitan Lima. *Sustainability* 13:2813. doi: 10.3390/su13052813
Hadjimichalis, C., and R. Hudson. 2014. Contemporary Crisis Across Europe and the Crisis of Regional Development Theories. *Regional Studies* 48:208–18. doi: 10.1080/00343404.2013.834044
Hall, S. 2011. Geographies of money and finance II. *Progress in Human Geography* 36 (3):403–11. doi: 10.1177/0309132511403889
Harvey, D. 1985. *The Urbanisation of Capital*. Oxford: Blackwell.

Harvey, D. 2003. *The New Imperialism*. Oxford: Oxford University Press.

Hassink, R. 2010. Regional resilience: a promising concept to explain differences in regional economic adaptability? *Cambridge Journal of Regions, Economy and Society* 3 (1):45–58. doi: 10.1093/cjres/rsp033

Henig D., and D. M. Knight. 2023. Polycrisis: prompts for an emerging worldview. *Anthropology Today* 39 (2):3–6. doi: 10.1111/1467-8322.12793

Ibert, O., M. Hess, J. Kleibert, F. Müller, and D. Power. 2019. Geographies of dissociation: Value creation, 'dark' places, and 'missing' links. *Dialogues in Human Geography* 9 (1):43–63. doi: 10.1177/2043820619831114

Ibert, O., G. Jackson, T. Theel, and L. Vogelgsang. 2021. Organizing Uncertainty as an Asset in Creative Collaboration: A Comparison of the Music and Pharmaceutical Industries. In *Organizing Creativity in the Innovation Journey*, ed. E. Schuessler, P. Cohendet, and S. Svejenova, 115–136. doi: 10.1108/S0733-558X20210000075010

Irwin, A., P. Simmons, and G. Walker. 1999. Faulty environments and risk reasoning: the local understanding of industrial hazards. *Environment and Planning A: Economy and Space* 31:1311–26. doi: 10.1068/a311311

Jessop, B., N. Brenner, and M. Jones. 2008. Theorising Sociospatial Relations. *Environment and Planning D: Society and Space* 26 (3):389–401. doi: 10.1068/d9107

Johnson, L. 2013. Catastrophe bonds and financial risk: Securing capital and rule through contingency. *Geoforum* 45:30–40. doi: 10.1016/j.geoforum.2012.04.003

Jones, M. and K. Ward. 2004. Capitalist Development and Crisis Theory: Towards a "Fourth Cut". *Antipode* 36: 497–511. doi:10.1111/j.1467-8330.2004.00428.x

Kitsos, A., and P. Bishop. 2018. Economic Resilience in Great Britain. The Crisis Impact and Its Determining Factors for Local Authority Districts. *Annals of Regional Science* 60 (2):329–47. doi: 10.1007/s00168-016-0797

Klagge, B. 2009. Financial Markets, Business Financing and the Recent Financial Crisis. *Zeitschrift für Wirtschaftsgeographie* 53 (1-2):1–13.

Kleibert, J. M. 2015. Global Production Networks, Offshore Services and the Branch-Plant Syndrome. *Regional Studies* 50 (12):1995–2009. doi: 10.1080/00343404.2015.1034671

Kleibert, J. M. 2021. Geographies of Marketization in Higher Education: Branch Campuses as Territorial and Symbolic Fixes. *Economic Geography* 97 (4):315–37. doi: 10.1080/00130095.2021.1933937

Knight, F. H. 1921. *Risk, Uncertainty, and Profit*. Boston: Houghton Mifflin.

Kruke, B. 2012. Societal safety and crisis management: Relevance for 22 July 2011. 22 July-Commission Paper, 7(12). 22 July Commission.

Kwak, D.-W., Y.-J. Seo, and R. Mason. 2018. Investigating the relationship between supply chain innovation, risk management capabilities and competitive advantage in global supply chains. *International Journal of Operations & Production Management* 38:2–21. doi: 10.1108/IJOPM-06-2015-0390

Lanari, N., D. Bek, J. Timms, and L. Simkin. 2021. In Whose Interests? Water Risk Mitigation Strategies Practiced by the Fruit Industry in South Africa's Western Cape. *Geoforum* 126:105–14. doi: 10.1016/j.geoforum.2021.07.025

Lawreniuk, S. 2020. Necrocapitalist networks: COVID-19 and the 'dark side' of economic geography. *Dialogues in Human Geography* 10 (2):199–202. doi: 10.1177/2043820620934927

Lawrence, M., S. Janzwood, and T. Homer-Dixon. 2022. 'What Is a Global Polycrisis?' Version 2.0. Discussion Paper 2022-4. Cascade Institute. Accessed February 7, 2024. https://cascadeinstitute.org/technical-paper/what-is-a-global-polycrisis/

Lee, R., G. J. Clark, J. Pollard, and A. Leyshon. 2009. The remit of financial geography—before and after the crisis. *Journal of Economic Geography* 9 (5):723–47. doi: 10.1093/jeg/lbp035

Lüder, L., and L. Kalvelage. 2023. Regional resilience and social-ecological systems: the impact of COVID-19 on community conservation in Namibia. *African Geographical Review*. Advance online publication. doi: 10.1080/19376812.2023.2276939

Lupton, D. 2013. *Risk*. 2nd ed. London: Routledge.
Martin, R. 2011. The Local Geographies of the Financial Crisis. From the Housing Bubble to Economic Recession and Beyond. *Journal of Economic Geography* 11 (4):587–618. doi: 10.1093/jeg/lbq024
Martin, R. L. 2018. Shocking Aspects of Regional Development: Towards an Economic Geography of Resilience. In *The New Oxford Handbook of Economic Geography*, ed. G. Clark, M. Gertler, M. P. Feldman, and D. Wójcik, 1st ed., 839–64. Oxford: Oxford University Press.
Martin, R., and P. Sunley. 2020. Regional economic resilience: evolution and evaluation. In *Handbook on Regional Economic Resilience*, ed. G. Bristow and A. Healy, 1st ed., 10–35. Cheltenham: Edward Elgar.
Martin, R., P. Sunley, B. Gardiner, and P. Tyler. 2016. How Regions React to Recessions: Resilience and the Role of Economic Structure. *Regional Studies* 50 (4):561–85. doi: 10.1080/00343404.2015.1136410
McConnell, A. 2003. Overview: Crisis Management, Influences, Responses and Evaluation. *Parliamentary Affairs* 56 (3):393–409. doi: 10.1093/parlij/gsg096
McCraw, T. K. 2009. *Prophet of Innovation: Joseph Schumpeter and Creative Destruction*. Cambridge: Harvard University Press.
Morsut, C., C. Kuran, B. I. Kruke, K. Orru, and S. Hansson. 2021. Linking resilience, vulnerability, social capital and risk awareness for crisis and disaster research. *Journal of Contingencies and Crisis Management* 30 (2):137–47. doi: 10.1111/1468-5973.12375
Mousavi, S., and G. Gigerenzer. 2014. Risk, uncertainty, and heuristics. *Journal of Business Research* 67 (8):1671–78. doi: 10.1016/j.jbusres.2014.02.013
Muellerleile, C., K. Strauss, B. Spigel, and T. P. Narins. 2014. Economic Geography and the Financial Crisis. Full Steam Ahead? *Professional Geographer* 66 (1):11–17. doi: 10.1080/00330124.2012.757819
Neise, T., T. López, and A. F. Angga Reksa. 2023. Rethinking Labour Risk in Global Production Networks: Resilience Strategies of Cruise Ship Workers in the Wake of the Covid-19 Pandemic. *Geoforum* 145:103842. doi: 10.1016/j.geoforum.2023.103842
Neise, T., J. Revilla Diez, and M. Garschagen. 2018. Firms as drivers of integrative adaptive regional development in the context of environmental hazards in developing countries and emerging economies – A conceptual framework. *Environment and Planning C: Politics and Space* 36 (8):1522–41. doi: 10.1177/2399654418771079
Neise, T., M. T. Sambodo, and J. Revilla Diez. 2021. Are micro-, small-and medium-sized enterprises willing to contribute to collective flood risk reduction? Scenario-based field experiments from Jakarta and Semarang, Indonesia. *Organization & Environment* 34 (2):219–42. doi: 10.1177/1086026619875435
November, V. 2008. Spatiality of Risk. *Environment and Planning A: Economy and Space* 40 (7):1523–27. doi: 10.1068/a4194
Omstedt, M. 2016. Reinforcing unevenness: post-crisis geography and the spatial selectivity of the state. *Regional Studies, Regional Science* 3 (1):99–113. doi: 10.1080/21681376.2015.1116959
Oßenbrügge, J. 2011. The Economic Crisis and the Reshaping of Geography. *DIE ERDE – Journal of the Geographical Society of Berlin* 142 (4):357–75.
Panitz, R., and J. Glückler. 2022. Relocation decisions in uncertain times: Brexit and financial services. *Economic Geography* 98 (2):119–44. doi: 10.1080/00130095.2021.2009336
Pažitka, V. and D. Wójcik. 2018. Cluster dynamics of financial centres in the UK: do connected firms grow faster? *Regional Studies* 53 (7):1017–28. doi: 10.1080/00343404.2018.1531116
Pidgeon, N. F., and J. Beattie. 1998. The Psychology of Risk and Uncertainty. In *Handbook of Environmental Risk Assessment and Management*, ed. P. Calow, 289–318. Oxford: Blackwell Science.

Pollard, J. S., J. Oldfield, S. Randalls, and J. E. Thornes. 2008. Firm finances, weather derivatives and geography. *Geoforum* 39 (2):616–24. doi: 10.1016/j.geoforum.2006.03.008

Rafiqui, P. S. 2009. Evolving economic landscapes: why new institutional economics matters for economic geography. *Journal of Economic Geography* 9 (3):329–53. doi: 10.1093/jeg/lbn050

Renn, O. 1990. Die gesellschaftliche Erfahrung und Bewertung von Risiken: Eine Ortsbestimmung [The social experience and assessment of risks: a localisation]. *Schweizerische Zeitschrift für Soziologie* 16:307–55.

Rodrigue, J.-P., T. Notteboom, and A. A. Pallis. 2011. The financialization of the port and terminal industry: revisiting risk and embeddedness. *Maritime Policy & Management* 38 (2):191–213. doi: 10.1080/03088839.2011.556675

Rohrmann, B. 2008. Risk Perception, Risk Attitude, Risk Communication, Risk Management: A Conceptual Appraisal. Keynote at the congress of The International Emergency Management Society TIEMS-2008, Prague.

Roitman, J. 2014. *Anti-crisis*. Durham: Duke University Press.

Roitman, J. 2021. Framing the crisis: COVID-19. Accessed February 20, 2024. https://papers.ssrn.com/sol3/papers.cfm?abstract_id=3952226

Rosenthal, U., A. Boin, and L. K. Comfort, eds. 2001. *Managing crises: Threats, dilemmas, opportunities*. Springfield, IL: Charles C. Thomas Publishers.

Rosenthal, U., M. T. Charles, and P. 't Hart. 1989. *Coping with crises: The management of disasters, riots and terrorism*. Springfield: Charles C. Thomas Publishers.

Schwabe, J. 2020. Risk and counter-strategies: The impact of electric mobility on German automotive suppliers. *Geoforum* 110:157–67. doi: 10.1016/j.geoforum.2020.02.011

Senanayake, N., and B. King. 2021. Geographies of uncertainty. *Geoforum* 123:129–35. doi: 10.1016/j.geoforum.2020.07.016

Simmie, J., and R. Martin. 2010. The economic resilience of regions: towards an evolutionary approach. *Cambridge Journal of Regions, Economy and Society* 3 (1):27–43. doi: 10.1093/cjres/rsp029

Slovic, P. 1992. Perception of risk: Reflections on the psychometric paradigm. In *Theories of risk*, ed. D. Golding, and S. Krimsky, 1st ed.,117–52. London: Praeger.

Sohns, F., and D. Wójcik. 2020. The impact of Brexit on London's entrepreneurial ecosystem: The case of the FinTech industry. *Environment and Planning A: Economy and Space* 52 (8):1539–59. doi: 10.1177/0308518X20925820

Sokol, M., and L. Pataccini. 2020. Winners and Losers in Coronavirus Times: Financialisation, Financial Chains and Emerging Economic Geographies of The Covid-19 Pandemic. *Tijdschrift voor economische en sociale geografie* 111 (3):401–15. doi: 10.1111/tesg.12433

Soroka, A., G. Bristow, M. Naim, and L. Purvis. 2020. Measuring regional business resilience. *Regional Studies* 54 (6):838–50. doi: 10.1080/00343404.2019.1652893.

Stanley, A. 2013. Natures of risk: Capital, rule, and production of difference. *Geoforum* 45:5–16. doi: 10.1016/j.geoforum.2012.06.010

Starr, R., J. Newfrock, and M. Delurey. 2003. *Enterprise Resilience: Managing Risk in the Networked Economy*. A strategy+business Reader, Booz Allen Hamilton Inc.

Strauss, K. 2008. Re-engaging with rationality in economic geography: behavioural approaches and the importance of context in decision-making. *Journal of Economic Geography* 8 (2):137–56. doi: 10.1093/jeg/lbm048

Strauss, K. 2009. Cognition, Context, and Multimethod Approaches to Economic Decision Making. *Environment and Planning A: Economy and Space* 41 (2):302–17. doi: 10.1068/a412

t'Hart, P., and A. R. Boin. 2001. Between crisis and normality: The long shadow of post-crisis politics. In *Managing Crises: Threats, Dilemmas, Opportunities*, ed. U. Rosenthal, A. R. Boin and L. Comfort, 1st ed., 28–46. Springfield: Charles C. Thomas.

Tooze, A. 2022. Welcome to the World of the Polycrisis. Accessed February 11, 2024. https://www.ft.com/content/498398e7-11b1-494b-9cd3-6d669dc3de33

Torrance, M. 2009. The Rise of a Global Infrastructure Market through Relational Investing. *Economic Geography* 85 (1):75–97. doi: 10.1111/j.1944-8287.2008.01004.x

Trippl, M., S. Fastenrath, and A. Isaksen. 2023. Rethinking regional economic resilience: Preconditions and processes shaping transformative resilience. *European Urban and Regional Studies*. Advance online publication. doi: 10.1177/09697764231172326

UNISDR. 2009. *UNISDR terminology for disaster risk reduction*. Geneva: United Nations.

Verfürth, P., T. Neise, M. Franz, and F. Sohns. 2022. The uneven effects of COVID-19 on the German restaurant and bar industry. *Erdkunde* (2):127–40. doi: 10.3112/erdkunde.2022.02.05

Vignoli, D., R. Guetto, G. Bazzani, E. Pirani, and A. Minello. 2020. A reflection on economic uncertainty and fertility in Europe: The narrative framework. *Genus* 76 (1):1–27. doi: 10.1186/s41118-020-00094-3

Völlers, P. 2023. Mitigating Extra-Firm Risk Environments – The Case of Turkish Firms in Germany. *Tijdschrift voor Economische en Sociale Geografie* 114 (3):234–51. doi: 10.1111/tesg.12557

Völlers, P., T. Neise, P. Verfürth, M. Franz, F. Bücken, and K. P. Schumacher. 2023. Revisiting Risk in The Global Production Network Approach 2.0 - Towards A Performative Risk Narrative Perspective. *Environment and Planning A: Economy and Space* 55 (8):1838–58. doi: 10.1177/0308518X231169288

Völlers, P., N. Yavan, and M. Franz. 2021. The reaction of foreign firms to institutional changes: The case of German direct investment in Turkey. *Applied Geography* 134:102503. doi: 10.1016/j.apgeog.2021.102503

Yeung, H. W.-C. 2022. *Interconnected Worlds. Global electronics and production networks in East Asia*. Stanford: Stanford University Press.

Yeung, H. W.-C. 2023. Troubling economic geography: New directions in the post-pandemic world. *Transactions of the Institute of British Geographers* 48 (4):672–80. doi: 10.1111/tran.12633

Wójcik, D. 2013. The Dark Side of NY–LON: Financial Centres and the Global Financial Crisis. *Urban Studies* 50:2736–52. doi: 10.1177/0042098012474513

Zehetmair, S. 2012. *Zur Kommunikation von Risiken: Eine Studie über soziale Systeme im Hochwasserrisikomanagement* [On the communication of risks: A study on social systems in flood risk management]. Wiesbaden: VS Verlag für Sozialwissenschaften.

3 Advancing Regional Economic Resilience Theorizing by Integrating Global Value Chain Resilience

Frank Jicha, Robert Hassink, and Han Chu

3.1 Introduction

Today's world is characterized time and again by crisis events. As a result, economic players in a wide variety of regional economies are constantly being challenged anew (Martin 2021). Depending on how well resilience is developed, regional actors come through a crisis differently (Hundt and Grün 2022). In economic geography, regional economic resilience is the most significant concept in understanding and explaining this question (Hassink and Gong 2020; Sutton et al. 2023). It looks at so-called determinants that affect the resilience of a regional economy. Although regional differences in the resilience of regional economies can be explained by regional determinants (Martin and Sunley 2020), the coronavirus pandemic has shown particularly clearly that there are also factors outside regional economies affecting the explanation of their resilience (Smorodinskaya, Katukov, and Malygin 2021). According to Yeung (2024, 1481) "… the world has experienced massive economic disruptions and (geo)political shifts taking place since the COVID-19 pandemic at the beginning of the 2020s. Localisation and regionalisation of production networks in the name of supply chain resilience and 'national' security have come to the forefront in considerations by many policy makers and business practitioners throughout the world" (see also Brakman, Garretsen, and van Witteloostuijn 2020).

However, these extra-regional factors have not yet been considered in the concept of regional economic resilience to a large extent. In line with recent similar, tentative attempts by Sutton et al. (2023) and Yeung (2024), this chapter aims to address this theoretical gap and hence to conceptually contribute to the economic geography concept of regional economic resilience. This will be done by drawing on the literature on global value chain resilience, which takes firm-related extra-regional factors into account. Although several concepts related to globalization, such as global production networks and global commodity chains, are discussed in the literature, in this chapter, we focus on global value chains, as they provide more theoretical and empirical literature related to resilience (Benfer et al. 2021; Yeung 2021). After presenting the core aspects of these two concepts in separate chapters, the subsequent discussion will

DOI: 10.4324/9781003457626-4

explicitly comment on the synthesis of these two concepts in order to identify research questions for future research on uncertainty, resilience, regional economies and value chains (Martin 2021).

3.2 Regional Economic Resilience

In economic geography, the concept of regional economic resilience examines how risk-prone and resilient actors in a regional economy react to shocks. On the one hand, the exact duration, frequency, origin, type and magnitude of shocks are examined (Gong et al. 2020; Sensier, Bristow, and Healy 2016). On the other hand, so-called determinants of resilience are studied (Di Caro and Fratesi 2018; Martin et al. 2016; Sutton et al. 2023). Only when a regional economy is exposed to a shock, it becomes clear how resilient it really is (Žítek and Klímová 2020). The aim is to explain why regions differ in their resilience in order to learn lessons about how to make regions and their businesses and workers less vulnerable to future shocks. This is all the more significant in today's environment of cascading crises (Martin 2021).

The regional industrial structure is one of the key determinants of regional economic resilience. Accordingly, a diverse economic structure makes a regional economy less vulnerable to shocks (portfolio effect), as it makes it more likely that some regional industries will not be affected by a shock. If there are also no input–output relationships to the affected industries, then they will be unaffected (Bishop 2019). In a more diverse regional economy, the recovery process tends to be faster, as there are more alternative opportunities to realign the economic base or development path of a regional economy to more resilient sectors (Balland, Rigby, and Boschma 2015; Boschma 2015; Martin and Sunley 2020).

Other important determinants include business culture, confidence and expectations (Martin and Sunley 2020), as well as local institutions, including norms and values (Huggins and Thompson 2015) and experiences to deal with past shocks and crises (institutional learning) (Bristow and Healy 2014).

These determinants of regional economic resilience can be seen as a product of the past or path dependence, since, for example, the business structure at the time of a shock has continuously evolved from the past (Diodato and Weterings 2015). In addition to current developments, this past development is therefore relevant for other determinants such as expectations and confidence, as this significantly influences the perception of a regional economy's economic strength. For example, a negative business climate may develop due to the lack of confidence among regional players, which means that in the event of a shock, urgently needed investments will not be made, even though this would be very important for the recovery process. On the contrary, in a regional economy that was prospering until the shock, regional actors may be so confident about the region's economic performance that investments continue to be made. This favors the recovery process accordingly (Martin and Sunley 2020). Debt structure and financial strength are also determinants that are closely related to local state support and national government economic policies (Martin and Sunley 2015).

The multitude of these different intra-regional determinants can therefore be used to make a comprehensive, well-founded assessment of the vulnerability, resilience and adaptive capacity or recovery process of a regional economy and the actors within it. Only by comprehensively analyzing all these determinants, the nature of resilience in each, specific, regional context can be fully grasped (Martin and Sunley 2015).

However, since in reality no regional economy exists completely isolated from other regional economies, the possible interactions between regional economies in terms of capital flows and labor shifts, with different degrees of intensity, need to be taken into account when analyzing regional economic resilience (see Figure 3.1). Figure 3.1 shows the different levels of resistance and recovery and combinations of both in reaction to a crisis. Moreover, it distinguishes between urban core regions, semi-core regions, semi-peripheral regions and peripheral regions. Peripheral regions might be in most cases structurally weak, but they might also contain some resilient enclaves. The same counts the other way around, high-resilient urban core regions might contain some pockets of low resilience. We call these differences in resilience within a regional economy, disparities in resilience. On the one hand, recovery processes and the degree of resistance can be observed in the regional economy as a whole, and on the other hand, in subregions of that region. Of course, subregions, in turn, are related to other subregions. The diversion of investment flows or labor to a more prosperous subregion then has a positive effect on the recovery process of a prosperous subregion, while at the same time worsening

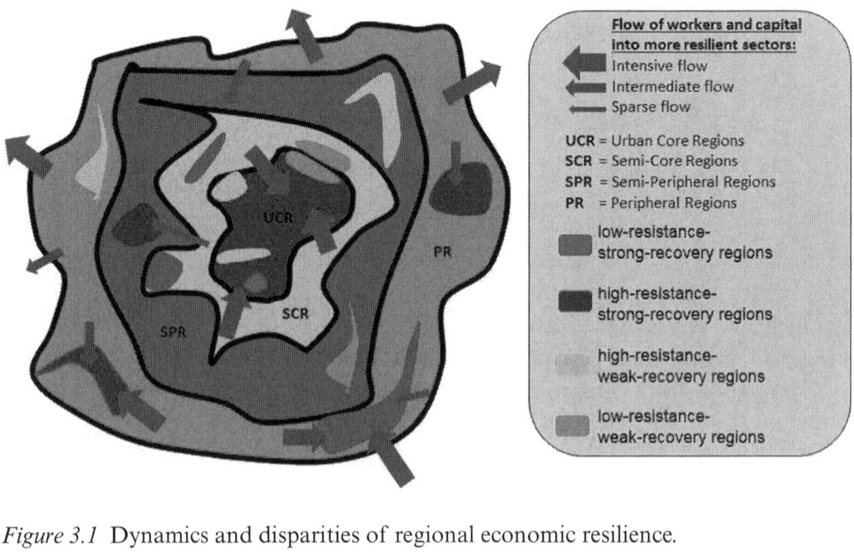

Figure 3.1 Dynamics and disparities of regional economic resilience.

Source: own design, based on Hundt and Grün 2022.

the initial conditions in the subregion that has become less attractive in the wake of a shock. However, these different developments within a regional economy cannot be explained solely by intra-regional flows of capital and labor; here, too, the argument of interdependencies with external regions affects the respective economic development. By increasing the resilience along a value chain, the resilience of all regions where this value chain is anchored is increased.

3.3 Extra-Regional Determinants of Regional Economic Resilience

As has been argued above, intra-regional determinants only partly explain differences in regional economic resilience. Therefore, extra-regional determinants have to be taken into account and they can be, broadly spoken, related to institutions and policies, but also to companies and their strategies and networks (see Figure 3.2).

National government economic policy comprises one important extra-regional determinant. In the concept of regional economic resilience, the state with its national institutions and its supportive policies is classified under intra-regional determinants, but in our view, national state support measures should be classified under extra-regional determinants (see Figure 3.2).

This also includes EU funding for companies in a regional economy (supranational level). Regions with better access to financial resources have more room to maneuver and also greater resilience if they use such capital to survive a crisis. Government support measures can be particularly significant for small and medium-sized enterprises (Brown et al. 2020), as they generally have fewer financial resources than large enterprises. Thus, the degree of activities of national and supranational policy authorities can affect the resilience of regions.

Moreover, several national and supranational policies affect investments and trade, such as import tariffs, quotas and export controls (Gereffi, Lim, and Lee 2021). These policies affect lead firms and suppliers, who can adapt and reconfigure their value chain if they are confronted with trade and investment restrictions (Gong et al. 2022), in order to minimize or prevent negative effects on their profits. Such a change in the value chain might affect resilience (Gong et al. 2022).

In a reaction to these policies, several switching strategies have been developed by firms, such as production switching, which involves relocating production facilities to other countries that are not affected by the restrictions. Classically, tariffs and import quotas are circumvented in this way (Gong et al. 2022). Market switching would be another alternative if products are sold in markets in other countries. For example, when previously exporting Chinese companies reorient themselves to the domestic market (Yang, Fu, and Li 2017). Last but not least, supplier switching should be mentioned, which is particularly interesting in the case of trade conflicts, as a change of suppliers to countries not affected by them makes it possible to deal with the restrictions (Gong et al. 2022). These different strategies show the importance of studying international firm strategies in relation to regional economic resilience. Although

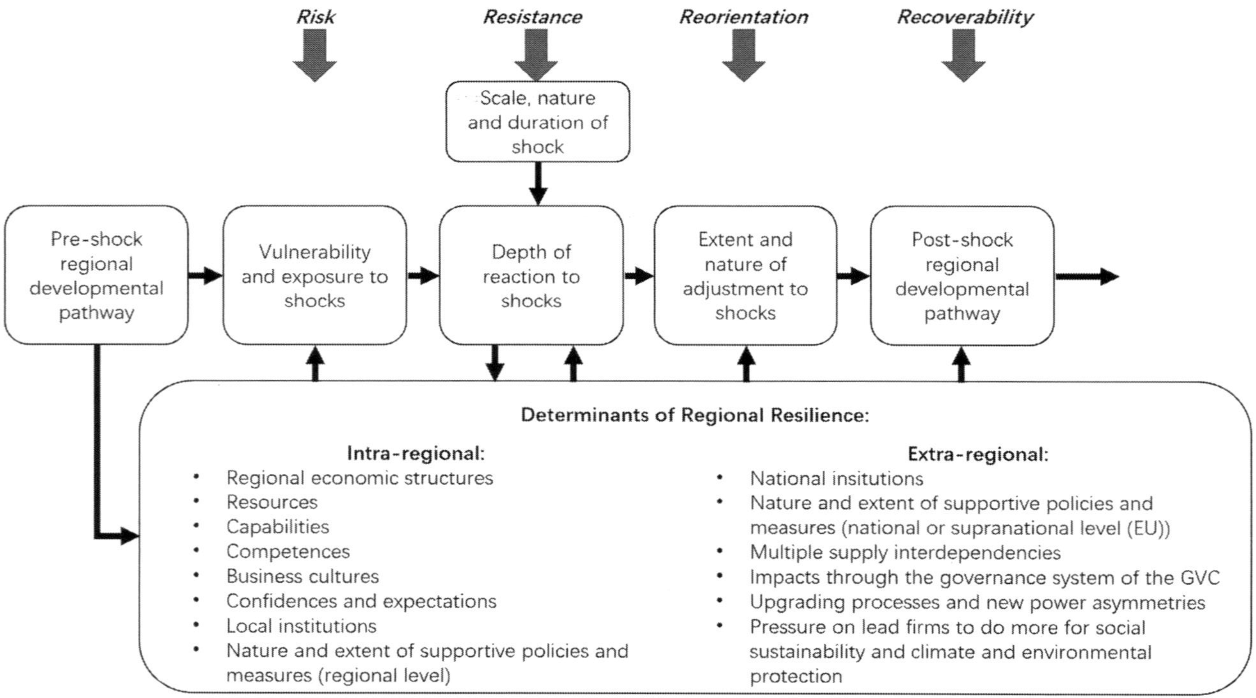

Figure 3.2 Expansion with extra-regional determinants.

Source: own design, based on Martin et al. 2016.

Martin and Sunley (2015) have discussed firm adjustment strategies in response to major shocks in the framework of regional economic resilience, little has been written about firm strategy in an international context in relation to regional economic resilience.

After discussing the intra-regional determinants of regional economic resilience in the previous section and reclassifying the aspect of policy impact as an extra-regional determinant, new firm-related extra-regional determinants will be presented in the following sub-sections.

3.3.1 Impact through the Governance System of Global Value Chains

Some regional economies are heavily impacted by decisions made by transnational corporations with headquarters located outside the regional or even national economy under consideration (Pla-Barber, Villar, and Narula 2021), which is an important firm-related extra-regional determinant affecting regional economic resilience (see Figure 3.2). One individual regional economy can be connected to several global value chains. Numerous supplier and sub-supplier firms from different countries cooperate in a network, in which a higher value is added to the repeatedly exported intermediate goods, whose transactions are coordinated by a lead firm (Cattaneo et al. 2013). Due to power asymmetries, lead firms have the authority to decide, for example, whether a value chain should be reconfigured (Barbieri et al. 2020). Classically, five types of global value chain governance structures have been identified, ranging from market, modular, relational, captive to hierarchy, with different degrees of explicit coordination and power asymmetry being the lowest with market and the highest with hierarchy (Gereffi, Humphrey, and Sturgeon 2005).

Value capture therefore is unevenly distributed along the value chain due to these power asymmetries; the lead firm usually retains most of it, but with relational or modular governance there is also the possibility of upgrading (Cattaneo et al. 2013). Lead firms in a value chain have various options for how exactly they want to design their value chain, ranging from sourcing within the company or with independent suppliers, as well as geographically, ranging from intra-regionally, domestically, to internationally (Cattaneo et al. 2013).

In general, lead firms prefer larger, more capable, globally operating, well-known firms as first-tier suppliers to small and medium-sized because larger suppliers are much more likely to be able to reliably fulfill larger orders once demand for the value chain products picks up again (Cattaneo, Gereffi, and Staritz 2010). Although this trend started before COVID-19, it became more pronounced during the pandemic. Some global value chains strive to become less dependent on certain network partners as well as on certain regions, which mainly refers to the overdependence on intermediates from China (Smorodinskaya, Katukov, and Malygin 2021). In this variant, to increase resilience, the geography of the value chain, that is, the number of suppliers at the global level, should be expanded. This so-called multiple-input sourcing consequently represents alternative procurement options at each production stage for intermediate goods in the event that certain regions and suppliers lack them due to a shock event. This can also apply to the procurement of raw materials by multiple suppliers (Smorodinskaya, Katukov, and Malygin 2021).

3.3.2 Multiple Supply Interdependencies

Multiple supply interdependencies are another relevant firm-related extra-regional determinant of regional economic resilience (see Figure 3.2). In many cases, hundreds of intermediary producers are used in the manufacturing of a product in the context of globalization. They are seen as a weak point that favors the vulnerability of global value chains (Smorodinskaya, Katukov, and Malygin 2021). A shock, can namely paralyze the entire value chain if significant suppliers cannot deliver, which is referred to as the ripple effect (Smorodinskaya, Katukov, and Malygin 2021). Therefore, shorter supply chains with fewer global partners should counteract the spread of shocks within the supplier network and thus increase the robustness of global value chains (Gölgeci, Yildiz, and Andersson 2020).

At the same time, the costs of coordination in shorter supply chains are lower, as are the costs of quality control. Shorter supply chains logically entail a lower risk for production sites, making such value chains less vulnerable (Pla-Barber, Villar, and Narula 2021). However, the complexity of the respective product must be taken into account, as this affects the number of suppliers. Depending on the product, global value chains differ in size and complexity and therefore also in their risk of supply disruptions (Gereffi 2020). Companies in a global value chain have no or only limited knowledge of suppliers outside their first-tier supplier. This is due to widespread subcontracting in many sectors worldwide. If such suppliers are unavailable due to a shock event, which is referred to as a hidden disruption, the global value chain is severely impacted (Smorodinskaya, Katukov, and Malygin 2021). The solution to hidden disruption is seen in digitalization as a resilience-promoting measure. Trace and tracking systems and big data analytics, for example, enable real-time data sharing and real-time coordination between all companies along the value chain. Radio frequency identification (RFID) chips also create transparency along the entire value chain for all inventories of, for example, intermediate goods from suppliers (Castañeda-Navarrete, Hauge, and López-Gómez 2021). Since digital technologies will create transparency regarding intermediate goods and gaps in the supply chains, lead firms see this as a key priority for more resilience. Post-pandemic digitalization has the potential to create a new generation of resilient global value chains in the long term, sharing information and data transparently and thereby collaborating effectively (Smorodinskaya, Katukov, and Malygin 2021). Other technologies include artificial intelligence, robotics and automation, big data, blockchain, 3D printing and machine learning, which can help to anticipate changes in demand and to change production as quickly as possible so that efficiency is increased (Dedrick and Kraemer 2005; Min 2019; Rogers et al. 2019). Cloud computing and the use of 5G are also expected to provide new data for this purpose, which will then be included in forecasts (Panwar, Pinkse, and De Marchi 2022).

3.3.3 Pressure on Lead Firms to Do More for Social Sustainability and Climate and Environmental Protection

There is constant pressure on lead firms to do more for social sustainability and climate and environmental protection, which is another determinant potentially affecting the resilience of regional economies (see Figure 3.2). Global value chains that are associated with lower greenhouse gas emissions or a more environmentally friendly global footprint are considered to be more resilient and better equipped for the future. Since consumers increasingly prefer sustainable products, selling more of them will contribute to more stable sales in the long term, which also has a beneficial effect on resilience in the event of a shock (Pla-Barber, Villar, and Narula 2021). Measures for more sustainability in value chains that lead to the so-called environmental upgrading of affected regions will be very important in the future in relation to climate change and climate protection and will also play an increasingly important role in the design of value chains, although short-term profit aspirations often run counter to these aims (Pla-Barber, Villar, and Narula 2021).

The regionalization of production networks is an important issue related to sustainability, because they have less impact on the environment due to shorter distances to transport intermediate goods (Panwar, Pinkse, and De Marchi 2022). Lead firms as controlling and steering authorities in global value chains can carry out corresponding reconfigurations (Cattaneo et al. 2013), such as so-called reshoring and streamlining and make value chains more regional. While reshoring means that, production activities are brought back from offshore locations to the home country of the transnational company, nearshoring specifically means switching to locations in a nearby country of the same macro-region (Barbieri et al. 2020), from which they benefit greatly, as new jobs are then created in the regional economy (Barbieri et al. 2020). The quality of the products produced as well as the generation of innovations can be increased if reshoring also involves R&D in addition to production facilities (Pla-Barber, Villar, and Narula 2021). Furthermore, it will also lead to intraregional trade increases, which also benefit the regional economy for example, due to regional service providers managing logistics, (Panwar, Pinkse, and De Marchi 2022). New, regional suppliers are incentivized to invest in their business through imminent regionalization, which increases efficiency (Pla-Barber, Villar, and Narula 2021).

3.3.4 Upgrading Processes and New Power Asymmetries

Upgrading processes and new power asymmetries are also related to regionalization processes, which got another boost during the recent COVID-19 pandemic, when decisions were taken to minimize risk exposure through reshoring

(Barbieri et al. 2020). At the beginning of the COVID-19 pandemic, when companies in a wide range of industries were really caught off guard, unexpected costs namely emerged, as lead firms had to manage complex offshore parts of supply chains (Pla-Barber, Villar, and Narula 2021). New regional suppliers potentially can benefit from learning opportunities when working with the lead firm and the rest of the companies in the global value chain, improving their skills and ultimately being able to perform higher value-added activities in the long run through this upgrading (Gölgeci, Yildiz, and Andersson 2020). Upgrading processes and new power asymmetries are therefore another firm-related extra-regional determinant (see Figure 3.2) because decisions from extra-regional headquarters of transnational corporations can largely change the economic structure and the power relations in the respective region.

Such an upgrading process from an original equipment manufacturer (OEM) to an original brand manufacturer (OBM) in the long run does not only apply to suppliers participating in overseas global production networks, but also to suppliers in the regional economy experiencing nearshoring (Panwar et al. 2022). Therefore, the pandemic potentially favored upgrading processes in regions with nearshoring (Pla-Barber, Villar, and Narula 2021). However, the global value chain governance of the respective lead firm also has to be taken into account, since depending on whether they use captive, relational or modular governance, there will be varying degrees of leeway for suppliers (Gereffi, Lim, and Lee 2021). After the coronavirus pandemic, lead firms that want to increase the resilience of their value chain to be prepared for the next crisis have changed their criteria for selecting suppliers. Preference has been given to more resilient suppliers fitting to the needs for a global value chain redesign, but such suppliers also demand more value capture, so that ultimately power dynamics in value chains are subject to change (Panwar, Pinkse, and De Marchi 2022). Regionalization of the value chain in certain affected industries will change the power asymmetries and governance structures. How exactly these shifts and potentially new forms of governance will unfold is subject to future empirical research of such cases (Panwar, Pinkse, and De Marchi 2022).

The establishment or acquisition of a center of excellence, in which R&D is carried out and new technological skills are researched, is considered extremely important for value creation and thus upgrading processes, from which the entire value chain can benefit (Panwar, Pinkse, and De Marchi 2022). In addition to the generation of local knowledge, external relationships are potentially a source of new knowledge for upgrading processes.

Moreover, smart sourcing strategies can help to increase the resilience of the value chain. They focus on encouraging innovation processes by locating the associated production facilities in the immediate vicinity of top universities or specialized clusters, thus gaining access to highly qualified labor markets (Smorodinskaya, Katukov, and Malygin 2021). Some lead firms also engage in R&D offshoring, not only focusing on industrialized countries, but also on developing such hotspots in emerging economies (Smorodinskaya, Katukov, and Malygin 2021).

Furthermore, a very important strategy for increasing resilience involves building more trust between interacting partner firms in the global value chain. This helps lead firms during a crisis to guarantee to receive their intermediate goods preferentially over other companies that also want to purchase intermediate goods from that supplier but are instead relegated to a waiting list (Pla-Barber, Villar, and Narula 2021). A lead firm can achieve this by counteracting power imbalances and granting suppliers more power and, logically, capturing value. This strongly strengthens the resilience of the value chain, but is again associated with costs for the lead firms. They strive for reducing uncertainty among key links in the value chain and for aligning the goals of involved companies so that supply relationships remain stable even in times of crisis (Pla-Barber, Villar, and Narula 2021).

Companies that do not implement adjustment measures run the risk of suffering strongly during future long-lasting or repeated shocks (Gölgeci, Yildiz, and Andersson 2020). Such players in a global value chain can, of course, hope for a timely end to a shock and then continue their business as usual. However, since today's world is characterized by uncertainty and risk, it is not advisable to operate global value chains in this way (Gölgeci, Yildiz, and Andersson 2020).

3.4 Introducing and Discussing Regional Value Chain Resilience

Intra-regional determinants of resilience allow statements about the resilience of a regional economy. However, intra-regional determinants are not the only criteria to be examined in the context of regional economic resilience; regional economies are linked to other regional economies through regional or global value chains, which, in the event of a severe shock, might negatively affect all interconnected regional economies if appropriate precautions have not been taken. Consequently, it seems logical that if the resilience of a regional economy or an industry rooted in a regional economy is to be determined, then the resilience of other regions connected to this region has to be examined as well. This means that all regions along an entire value chain would have to be examined in order to make a comprehensive assessment of a regional economy's resilience. The previous focus of regional economic resilience mainly on one region and intra-regional determinants is therefore insufficient. Hence, we propose to use the term *regional value chain resilience* as a sub-category of the concept of regional economic resilience that is particularly relevant in those regional economies that are strongly embedded in global value chains. Regional value chain resilience can be regarded as the ability to cope with shocks and crises of that part of the regional economy that is strongly embedded in global value chains and hence is also influenced by resilience of the global value chain and its related regions. As has been pointed out in Section 3.3, regional value chain resilience is influenced by several firm-related extra-regional determinants related to value chain resilience, that is the governance system of the global value chain, multiple supply interdependencies, pressure related to sustainability issues, as well upgrading processes and power asymmetries.

Therefore, in our view, future empirical studies with a stronger value chain perspective would be necessary to comprehensively assess the resilience of a regional economy. We therefore very much follow Yeung's (2024) recent suggestion to think increasingly of a combination of regional and network resilience. According to him "the post-pandemic scalar shift from the global 'back again' to the regional in relation to geopolitical imperatives and risk mitigation will necessitate the rethinking of how production networks can be regionalised in ways that make them more robust and resilient to external shocks and risks. As such, *network and regional resiliences* are critical agendas …" (Yeung 2024, 1487, italics in original).

Moreover, regional value chain resilience is particularly relevant for many regional clusters (Chu and Hassink 2022), and therefore there is an interesting connection to cluster resilience (Crespo, Suire, and Vicente 2014; Martin and Sunley 2011; Chu and Gong 2024), which can be regarded as an additional sub-category of the concept of regional economic resilience.

We see much potential in studying regional value chain resilience as a sub-category of regional economic resilience in future economic geography research, but realize that the concept is not relevant to each and every regional economy. It is particularly in regional economies with strong industrial specialization and clusters which are embedded in global value chains that we see this potential. In our view, three key topics should be emphasized in future research on regional value chain resilience as a sub-category of regional economic resilience.

First, the role of the lead firms in global value chains and their decisions are a key factor affecting several extra-regional determinants of the resilience of some regional economies. At the lead firm's headquarters, decisions are made about reshoring, about whether the entire value chain should become shorter, more regional and sustainable, as well as about the selection of more or less resilient suppliers. The lead firms do not only have large decision-making power, but also claim the largest share in the value capture of the revenues generated. Lead firms have various value chain design options, which affect the resilience of a regional economy in different ways. They also decide how to deal with a shock at the respective production site or whether the production operation has to be stopped. Moreover, they also take important decisions on stocks, inventories, and just-in-time or just-in-case production, affecting the vulnerability of the value chain (Gereffi 2020; Gölgeci, Yildiz, and Andersson 2020). They might make costly investments to increase resilience by having larger stocks, foregoing the benefits of cost-saving just-in-time production. Therefore, a key question for future research is how best to increase resilience in the value chain both in the short term, but also in the long term, so that it can be truly said to be sustainable (Pananond, Gereffi, and Pedersen 2020) and how that is related to the resilience of regions strongly linked to these chains. Alternatively, lead firms focus resilience-building measures only on the parts of the value chain where critical inputs are involved (Pananond, Gereffi, and Pedersen 2020). For policy-makers in regional economies, it is key to maintain the resilient parts of the value chains in the region, which will positively affect the regional economy's success and resilience in the long term.

Second, differences between industries, particularly concerning their potential for regionalizing and reshoring, are another key topic that should be emphasized in future research on regional value chain resilience as a sub-category of regional economic resilience. Certain industries, such as the automotive industry, advanced technologies, mechanical engineering, electronics and high service sectors, which are characterized by medium- and high innovation intensity and customization, are prone to be affected by regionalization of the value chain. Moreover, in these industries, switching back to offshoring after a regionalization process is unlikely, even in the long term, because supply reliability or quality standards are in these cases just as important as labor costs (Barbieri et al. 2020). Such value chains will become shorter, more regional and add more value at each stage of the value chain (Gölgeci, Yildiz, and Andersson 2020). More research is needed on how relocation, reshoring and nearshoring, and upgrading processes in these industries affect regional value chain resilience. In other industries, such as the clothing industry in the low-cost sector, reshoring does make less sense and hence alternative measures have to be chosen to increase resilience (Pla-Barber, Villar, and Narula 2021). Moreover, mining, energy or agricultural sectors will not be affected by regionalization of the value chain because their key inputs are location-bound (Pla-Barber, Villar, and Narula 2021).

The third key topic arises from the differences between resilience, robustness and the paradox within regional value chain resilience. As mentioned earlier, sometimes "multiple input sourcing" can provide alternative options for disruptions caused by shocks, which actually enhances the resilience of a region (Smorodinskaya, Katukov, and Malygin 2021). However, in some cases, when supply chain systems are shorter and simpler, mostly relying on partners within rather than outside the region, this instead facilitates improved robustness (Gölgeci, Yildiz, and Andersson 2020). We need to determine, based on different regional geographical backgrounds and industrial structural differences, which type of capability should be prioritized in a region in response to various types of crises, in order to enable the region to successfully navigate through crises and shocks.

3.5 Conclusion

Overall, we see the need to extend the regional economic resilience concept in economic geography with stronger research on value chain resilience, and proposed regional value chain resilience as a sub-category of regional economic resilience. This concept not only reveals the significant impact of external factors on regional economic resilience, but also emphasizes the importance of adjusting the region's position and role in the value chain according to the specific global network dynamics of the region. As our analysis has highlighted, the decision between diversifying or simplifying supply and procurement networks can significantly affect a region's ability to withstand economic shocks. Particularly, the recent COVID-19 pandemic, as well as trade wars, have shown that the resilience of some regional economies is strongly related to global value chain disruptions, which therefore should be dealt with more prominently in both regional economic resilience theorizing, as well as future empirical research. Sometimes, regions with diversified networks are able to mitigate some adverse effects by leveraging alternative supplies. Conversely, some regions might benefit more from shorter, more localized supply chains. The cascading crises as well the unique character of each of them individually, in combination with the unique character of each regional economy, has strongly increased the complexity of measuring, analyzing and understanding regional economic resilience, as well as explaining why some regional economies are more resilient than others. Moreover, the increased complexity and the particularities of crises and regions will make it also challenging to distill generalizable results from empirical studies (Gong and Hassink 2020; Cox and Evenhuis 2020). In light of these insights, future research should aim to develop more refined theoretical frameworks and measurement tools to assess the resilience of regional value chains. Ultimately, this research contributes to a deeper understanding of how regions can better cope with the complexities of global interdependence by focusing on strategic adjustments tailored to specific regional and industry contexts, thereby enhancing their own economic resilience.

Acknowledgments

This work has been sponsored by the German Research Foundation (DFG) in the framework of the project: Exploring the impact of COVID-19 on regional resilience and disparities in Germany and China (project number: 458455447).

References

Balland, P. A., D. Rigby, and R. Boschma. 2015. The technological resilience of US cities. *Cambridge Journal of Regions, Economy and Society* 8 (2):167–84. https://doi.org/10.1093/cjres/rsv007

Barbieri, P., A. Boffelli, S. Elia, L. Fratocchi, M. Kalchschmidt, and D. Samson. 2020. What can we learn about reshoring after Covid-19? *Operations Management Research* 13:131–36. https://doi.org/10.1007/s12063-020-00160-1

Benfer, M., B. Verhaelen, S. Peukert, and G. Lanza. 2021. Resilience Measures in Global Production Networks. *Die Unternehmung* 75 (4):491–520. https://doi.org/10.5771/0042-059X-2021-4-491

Bishop, P. 2019. Knowledge diversity and entrepreneurship following an economic crisis: an empirical study of regional resilience in Great Britain. *Entrepreneurship & Regional Development* 31(5-6):496–515. https://doi.org/10.1080/08985626.2018.1541595

Boschma, R. 2015. Towards an Evolutionary Perspective on Regional Resilience. *Regional Studies* 49(5):733–751. https://doi.org/10.1080/00343404.2014.959481

Brakman, S., H. Garretsen, and A. van Witteloostuijn. 2020. The turn from just-in-time to just-in-case globalization in and after times of COVID-19: An essay on the risk re-appraisal of borders and buffers. *Social Sciences & Humanities Open* 2 (1):100034. https://doi.org/10.1016/j.ssaho.2020.100034

Bristow, G., and A. Healy. 2014. Regional Resilience: An Agency Perspective. *Regional Studies* 48 (5):923–35. https://doi.org/10.1080/00343404.2013.854879

Brown, R., R. V. Kalafsky, S. Mawson, and L. Davies. 2020. Shocks, uncertainty and regional resilience: The case of Brexit and Scottish SMEs. *Local Economy* 35 (7):655–75. https://doi.org/10.1177/0269094220979261

Castañeda-Navarrete, J., J. Hauge, and C. López-Gómez. 2021. COVID-19's impacts on global value chains, as seen in the apparel industry. *Development Policy Review* 39 (6):953–70. https://doi.org/10.1111/dpr.12539

Cattaneo, O., G. Gereffi, S. Miroudot, and D. Taglioni. 2013. Joining, upgrading and being competitive in global value chains: A strategic framework. World Bank Policy Research Working Paper. 6406.

Cattaneo, O., G. Gereffi, and C. Staritz. 2010. Global value chains in a postcrisis world: resilience, consolidation, and shifting end markets. In *Global value chains in a postcrisis world: a development perspective*. ed. O. Cattaneo, G. Gereffi, and C. Staritz, 3–20. Washington. https://doi.org/10.1596/978-0-8213-8499-2

Chu, H., and H. Gong. 2024. All Roads Lead to… Resilience?! Unravelling Cluster Resilience in the Polycrisis Era. Unpublished manuscript.

Chu, H., and R. Hassink. 2022. Regional Clusters. In *Oxford Bibliographies in Urban Studies*, ed. R. Dilworth. New York: Oxford University Press. https://doi.org/10.1093/OBO/9780190922481-0058

Cox, K. R., and E. Evenhuis. 2020. Theorising in urban and regional studies: Negotiating generalisation and particularity. *Cambridge Journal of Regions, Economy and Society* 13 (3):425–42. https://doi.org/10.1093/cjres/rsaa036

Crespo, J., R. Suire, and J. Vicente. 2014. Lock-in or lock-out? How structural properties of knowledge networks affect regional resilience. *Journal of Economic Geography* 14 (1):199–219. https://doi.org/10.1093/jeg/lbt006

Dedrick, J., and K. L. Kraemer. 2005. The impacts of IT on firm and industry structure: The personal computer industry. *California Management Review* 47 (3):122–42. https://doi.org/10.2307/41166309

Di Caro, P., and U. Fratesi. 2018. Regional determinants of economic resilience. *The Annals of Regional Science* 60:235–40. https://doi.org/10.1007/s00168-017-0858-x

Diodato, D., and A. B. Weterings. 2015. The resilience of regional labour markets to economic shocks: Exploring the role of interactions among firms and workers. *Journal of Economic Geography* 15 (4):723–42. https://doi.org/10.1093/jeg/lbu030

Gereffi, G. 2020. What does the COVID-19 pandemic teach us about global value chains? The case of medical supplies. *Journal of International Business Policy* 3 (3):287–301. https://doi.org/10.1057/s42214-020-00062-w

Gereffi, G., J. Humphrey, and T. Sturgeon. 2005. The governance of global value chains. *Review of International Political Economy* 12 (1):78–104. https://doi.org/10.1080/09692290500049805

Gereffi, G., H. C. Lim, and J. Lee. 2021. Trade policies, firm strategies, and adaptive reconfigurations of global value chains. *Journal of International Business Policy* 4 (4):506–22. https://doi.org/10.1057/s42214-021-00102-z

Gölgeci, I., H. E. Yildiz, and U. R. Andersson. 2020. The rising tensions between efficiency and resilience in global value chains in the post-COVID-19 world. *Transnational Corporations Journal* 27 (2):127–41. https://doi.org/10.18356/99b1410f-en

Gong, H., and R. Hassink. 2020. Context sensitivity and economic-geographic (re) theorising. *Cambridge Journal of Regions, Economy and Society* 13 (3):475–90. https://doi.org/10.1093/cjres/rsaa021

Gong, H., R. Hassink, C. Foster, M. Hess, and H. Garretsen. 2022. Globalisation in reverse? Reconfiguring the geographies of value chains and production networks. *Cambridge Journal of Regions, Economy and Society* 15 (2):165–81. https://doi.org/10.1093/cjres/rsac012

Gong, H., R. Hassink, J. Tan, and D. Huang. 2020. Regional Resilience in Times of a Pandemic Crisis: The Case of COVID-19 in China. *Tijdschrift Voor Economische En Sociale Geografie* 111 (3):497–512. https://doi.org/10.1111/tesg.12447

Hassink, R., and H. Gong. 2020. Regional Resilience. In *International Encyclopedia of Human Geography*, ed. A. Kobayashi, 2nd ed., 351–55. Elsevier. https://doi.org/10.1016/B978-0-08-102295-5.10126-X

Huggins, R., and P. Thompson. 2015. Local entrepreneurial resilience and culture: the role of social values in fostering economic recovery. *Cambridge Journal of Regions, Economy and Society* 8 (2):313–30. https://doi.org/10.1093/cjres/rsu035

Hundt, C., and L. Grün. 2022. Resilience and specialization–How German regions weathered the Great Recession. *ZFW–Advances in Economic Geography* 66 (2): 96–110. https://doi.org/10.1515/zfw-2021-0014

Martin, R. 2021. Rebuilding the economy from the Covid crisis: time to rethink regional studies? *Regional Studies, Regional Science* 8 (1):143–61. https://doi.org/10.1080/21681376.2021.1919191

Martin, R., and P. Sunley. 2011. Conceptualizing cluster evolution: beyond the life cycle model? *Regional Studies* 45 (10):1299–1318. https://doi.org/10.1080/00343404.2011.622263

Martin, R., and P. Sunley. 2015. On the notion of regional economic resilience: conceptualization and explanation. *Journal of Economic Geography* 15 (1):1–42. https://doi.org/10.1093/jeg/lbu015

Martin, R., and P. Sunley. 2020. Regional economic resilience: evolution and evaluation. In *Handbook on Regional Economic Resilience*, ed. G. Bristow, and A. Healy, 10–35. Cheltenham: Edward Elgar. https://doi.org/10.4337/9781785360862.00007

Martin, R., P. Sunley, B. Gardiner, and P. Tyler. 2016. How regions react to recessions: Resilience and the role of economic structure. *Regional Studies* 50 (4):561–85. https://doi.org/10.1080/00343404.2015.1136410

Min, H. 2019. Blockchain technology for enhancing supply chain resilience. *Business Horizons* 62 (1):35–45. https://doi.org/10.1016/j.bushor.2018.08.012

Pananond, P., G. Gereffi, and T. Pedersen. 2020. An integrative typology of global strategy and global value chains: The management and organization of cross-border activities. *Global Strategy Journal* 10 (3):421–43. https://doi.org/10.1002/gsj.1388

Panwar, R., J. Pinkse, and V. De Marchi. 2022. The future of global supply chains in a post-COVID-19 world. *California Management Review* 64 (2):5–23. https://doi.org/10.1177/00081256211073355

Pla-Barber, J., C. Villar, and R. Narula. 2021. Governance of global value chains after the Covid-19 pandemic: a new wave of regionalization. *BRQ Business Research Quarterly* 24 (3):204–13. https://doi.org/10.1177/23409444211020761

Rogers, W. P., M. M. Kahraman, F. A. Drews, K. Powell, J. M. Haight, Y. Wang, K. Baxla, and M. Sobalkar. 2019. Automation in the mining industry: Review of technology, systems, human factors, and political risk. *Mining, Metallurgy & Exploration* 36:607–31. https://doi.org/10.1007/s42461-019-0094-2

Sensier, M., G. Bristow, and A. Healy. 2016. Measuring regional economic resilience across Europe: Operationalizing a complex concept. *Spatial Economic Analysis* 11 (2):128–51. https://doi.org/10.1080/17421772.2016.1129435

Smorodinskaya, N. V., D. D. Katukov, and V. E. Malygin. 2021. Global value chains in the age of uncertainty: advantages, vulnerabilities, and ways for enhancing resilience. *Baltic Region* 13 (3):78–107. https://doi.org/10.5922/2079-8555-2021-3-5

Sutton, J., A. Arcidiacono, G. Torrisi, and R. N. Arku. 2023. Regional economic resilience: A scoping review. *Progress in Human Geography* 47 (4):500–32. https://doi.org/10.1177/03091325231174183

Yang, C., T. Fu, and L. Li. 2017. Emerging adaptation of local clusters in China in a shifting global economy: evidence from the furniture cluster in Houjie Town, Dongguan. *Growth and Change* 48 (2):214–32. https://doi.org/10.1111/grow.12191

Yeung, H. W. C. 2021. The trouble with global production networks. *Environment and Planning A: Economy and Space* 53 (2):428–38. https://doi.org/10.1177/0308518X20972720

Yeung, H. W. C. 2024. From regional to global and back again? A future agenda for regional evolution and (de)globalised production networks in regional studies. *Regional Studies* 58 (7):1480–1491. https://doi.org/10.1080/00343404.2024.2316856

Žítek, V., and V. Klímová. 2020. Regional resilience redefinition: Postpandemic challenge. Scientific Papers of the University of Pardubice. Series D. *Faculty of Economics and Administration* 28 (4):1158. https://doi.org/10.46585/sp28041158

4 Jenga Capitalism and the Cyber-Energy-Production Plexus

New Forms of Risk and Uncertainty That Are Reshaping Economic Geography

John R. Bryson

4.1 Introduction

All economic actions and activities are saturated with risks and uncertainties. In economics, the distinction between risk and uncertainty can be traced back to Knight (1921); risk has an unknown outcome, but the probability distribution is known whilst uncertainty comes with unknown outcomes and probabilities. There are many different types of risk with some being directly or indirectly related to human activity and other risks being the outcome of natural processes. One of the challenges is that "so long as risks are not recognized scientifically, they do not exist" (Beck 1992, 71). This is an important point in that a risk must be recognized for it to be defined as a risk whilst uncertainties are a much broader category of events, processes and activities that could potentially have negative or positive impacts. Human activity also increases the probability of environmental and other types of risks, and this includes climate change.

Economic activity has always been exposed to all types of risks, uncertainties, and crisis. Some risks are predictable or recurrent events but of unknown timing or duration, for example, recessions, inflationary pressures, extreme weather events impacting on harvests, epidemics or pandemics and wars. What differentiates this century from earlier times is that governments, companies, and individuals have enhanced ability to forecast or anticipate risks. More data is available from satellites and sensors, and this includes Big Data. This enhances the ability of some companies and governments to invest in resources and operational procedures that are intended to enhance certainty or reduce the impacts of risks of all types. However, this process of risk anticipation requires decisions regarding the prioritization of investments intended to minimize the impacts of risks that might eventually result in events that incur costs. This century, however, is one in which there has been an exponential increase in societal and individual exposure to risk and uncertainty. What is occurring is the emergence of a new period of capitalism that is best defined as *Jenga Capitalism*. Jenga Capitalism combines the noun capitalism with the proper noun Jenga. Jenga is a game in which individual players take turns to remove

DOI: 10.4324/9781003457626-5

54 wooden blocks from a stack or tower with the loser being the player whose actions cause the tower to fail. Each block that is removed is placed on the top of the structure with the tower becoming increasingly precarious. One driver behind the emergence of Jenga Capitalism is that the socioeconomy is in an on-going process of becoming too dependent on interdependent systems with an excellent example being the Internet of Things (IoT). With the IoT connected electronic devices' functionality may be completely dependent on being connected to a network. The outcome of enhanced system interdependency is that failure in one part of a system can ricochet throughout the system resulting in unexpected consequences.

Risks may not result in damage, but uncertainty may prevent investment that will then impact on innovation, productivity, and competitiveness. A business may be prepared for the impacts of a crisis or may be surprised by an event. Nevertheless, it is important to remember that any one crisis will be simultaneously experienced with a series of on-going crises of different types, durations, and intensity (Bryson et al. 2024). Thus, the COVID-19 pandemic was a crisis whose impacts were layered upon on-going crises, and new crises emerged during the pandemic. Thus, all businesses are in a continual state of adapting to crises and in activities that are intended to enhance certainty and reduce uncertainty. New types of risk have emerged with technological innovation and enhanced competition. Nevertheless, it is important to appreciate that throughout history all businesses have had to engage in some form of risk planning that is intended to reduce uncertainty. A good example are the businesses owned by Francesco di Marco Datini, otherwise known as the Merchant of Prato. Francesco was an Italian merchant who was born in Prato in around 1335 and died in 1410. Underdeveloped logistics enhanced the risks related to Francesco's business. Thus, in the cloth trade at this time a period of "no less than three and a half years were required" "between the first expenditure of capital and any returns" (Origo 1988, 75). It has been calculated that the profit margin was no more than 8.92% and this was a small return given all these "imponderable hazards – from shipwreck, piracy, land-robbers, dishonest agents, closed ports, pestilence, and deterioration of wares" (Origo 1988, 76). All businesses require access to information to inform a risk assessment process. For Francesco, this involved frequent correspondence with trading centers located across Europe "with all these letters having one thing in common: every event they report – a battle or a truce, a rumour of pestilence, famine or flood, a Pope's election or a Prince's marriage – is noted only with a view to its effect upon trade" (Origo 1988, 83).

The Merchant of Prato is an excellent example of an international business that was saturated with risks and uncertainty. This was especially the case given the then nature of the available communications and logistics networks. In 2024, information can be exchanged between multiple points on this planet immediately, but technological innovations have created new forms of risk that have also enhanced uncertainty. This chapter is a call for economic geography to develop a more sophisticated approach to conceptualizing risk and uncertainty. The chapter begins with a discussion of the risk society. This then shifts to a critique of Beck's risk society thesis (Beck 1992) as a new epoch of capitalism has emerged that comes with a very different risk profile to that of Beck's new modernity. The emergence of Jenga Capitalism is explored and theorized, before exploring the geographical implications of this new configuration of capitalist relationships. Finally, the chapter outlines a new approach to conceptualizing risk in economic geography.

4.2 The Risk Society

Since the publication of Ulrich Beck's Risikogesellschaft in Germany in 1986, and its subsequent translation in 1992 as *Risk Society*, Beck's work has continued to stimulate debate on risk and modernity. Beck's risk society thesis led to a debate regarding the changing nature of risks and related debates on resilience and how risks can be minimized. Central to Beck's thesis is the argument that new modernity transforms risk. This transformation undermines approaches to calculating risk; the outcomes of a risk that is actualized transitions from producing local impacts to catastrophic events. Beck differentiates between calculable and non-calculable threats and notes that "under the surface of risk calculation new kinds of *industrialized, decision-produced incalculabilities and threats* are spreading within the globalization of high-risk industries, whether for warfare or welfare purposes" (Beck 1992, 22, italics in the original). The incalculability of the consequences of these risks is a core feature of Beck's argument.

The risk society thesis is part of a major criticism of social science as it includes a call to "break out of the iron cage of conventional and orthodox social science and politics" and the need for a "new sociological imagination" (Beck 2000, 212). There are five important caveats to make regarding Beck's risk society thesis. First, the identification of some form of risk does not necessarily imply that any damage will be incurred. Risk must not be confused with the impacts experienced by a society when a risk results in some degree of destruction. Nevertheless, a distinction must be made between direct, indirect, and induced impacts of risk events. Direct impacts are forms of destruction that can be directly attributed to a risk. These include the destruction of property experienced during flooding. Indirect impacts include, for example, longer-term negative impacts on economic activity resulting from the impacts of a risk event. Risks that are actualized and result in damage may act as a discontinuity that disrupts economic activity and this disruption has a long-term tempering impact on an economy. Induced impacts of risks result from activities intended to mitigate against perceived risks. Risk mitigation involves capital and revenue expenditure. Such expenditure might have prevented any damage from incurring once a risk is actualized or might have reduced the negative impacts. Nevertheless, this type of expenditure might never have been required as the risk that has been identified might never occur. There are then opportunity costs related to expenditure intended to mitigate risks. These opportunity costs emerge as expenditure is allocated to risk mitigation that could be invested to support innovation or activities that could result in activities that supported additional wealth generation initiatives and/or societal benefits. There is a real danger in assuming that expenditure intended to mitigate risks results in beneficial impacts. Nevertheless, such expenditure will still support employment and will create local multipliers.

To Beck "the concept of risk reverses the relationship of past, present and future" as "the past loses its power to determine the present" (Beck 2000, 214). The argument is that calculations regarding future risks determine present-day actions with political, media and academic debate focused on exploring events that might happen and can only be avoided by altering socio-economic activities. There are many problems with this type of analysis. On the one hand, people tend to live their lives in the present and are influenced by prior experiences rather than by debates regarding different types of future incalculable risks. This suggests that this reversal of the past, present, and future does not apply to all people or organizations, but only to those engaged in debates about incalculable risks. On the other hand, for governments and companies calculating the level of investment required to respond to some type of future risk is a complicated activity. This explains the level of funding allocated to mitigate and adapt to climate change. The difficulty is that economists apply a 'discount rate' to future costs and benefits. This technique is based on assessing the time value of money and the risks related to future benefits and costs. This is a complex process based on "discounting costs incurred in later years by an appropriate discount rate compared with costs incurred in earlier years" (Letwin 2020, 171). This process will recommend "taking the course of action in question only if (a) the discounted benefits exceed the discounted costs and (b) the extent of the net benefit is sufficiently great to rank the course of action ahead of other net-beneficial actions in other areas in which one could invest the money" (Letwin 2020, 171). The difficulty is that it is nearly impossible to calculate the probability of a risk occurring in a particular year, especially if these are recognized as being incalculable events. The outcome of the application of the discount rates approach to cost-benefit analysis is that low-probability and distant events are heavily discounted with the result being that the calculation places very little weight on risks that might happen at some time in the future. For governments and companies, the calculation of discount rates and their application to risk mitigation strategies undermines the emphasis Beck places on calculations of future events determining present actions.

The risk society thesis contains three core notions: organized irresponsibility, relations of definition and social explosiveness of hazards (Beck 2000, 224). These concepts underpin the emphasis that Beck places on the need to theorize the risk society. First, the concept of *organized irresponsibility* highlights that the risk society is characterized by perceived and potential environmental degradation and yet no one is accountable. Thus, industrialization, and consumer behavior, is organized and regulated, but the collective outcome of many individual decisions and investments enhances risks and results in incalculable risks. An excellent example of this process is provided in a survey of

26,358 people on climate change undertaken by the European Commission in 2023. Respondents were asked about "which actors within the EU (from a list of six) they think are responsible for tackling climate change" with multiple responses being possible (European Union 2023, 27). Only 35% of respondents stated that they themselves were personally responsible whilst 35% stated that responsibility lies with regional and local authorities. Only 6% noted that all six actors were responsible. The risks related to climate change result from the actions of all actors – governments, companies, organizations, and all individuals. Each of these actors is regulated and each, however, engages in activities that enhance the risks related to climate change. Thus, each actor considers that they are acting responsibility whilst collectively the outcome is organized irresponsibility.

Second, *relations of definition* "include the specific rules, institutions and capacities that structure the identification and assessment of risk in a specific cultural context" (Beck 2000, 224). These are processes and techniques that lead to the identification and definition of risk. A good example is Russia's war with Ukraine that commenced in 2014 with Russia's first invasion of Ukraine that resulted in the illegal annexation of Crimea, and this then led to the second invasion of Ukraine on 24 February 2022. This invasion was justified by Russia on the grounds that Ukraine was a threat to Russia's security. There is no evidence to support the assessment that Ukraine was a real security threat to Russia. Thus, this is an example of risk being defined for political purposes rather than the identification of any actual risk that could lead to actions that would damage Russia. There is an interesting paradox here in that Russia's invasion of Ukraine has damaged Russia to an intensity and degree that cannot be justified by any so-called gains. All Russian citizens will continue to experience negative impacts that come from this war for decades.

Third, Beck uses the metaphor of the *social explosiveness of hazards* to highlight the ways in which risks become real or become acknowledged as potential risks that require mitigation and adaptation. In this account, the emphasis is placed on the ways in which the identification of new hazards as socially constructed results in these risks emerging as a form of "quasi-subjects" that are a "powerful, uncontrollable 'actor' that delegitimates and destabilizes state institutions" (Beck 2000, 225). This type of destabilization, or delegitimation, process has not occurred to any great extent as risks are identified and debated. Part of the problem is that individual and collective action continues to support activities that collectively result in organized irresponsibility. Central to this process is the failure of cost-benefit analysis to take societal risks seriously and for discount rates to be calculated that fail to recognize the extent of the changes that are required to mitigate risks. For the individual, life continues very much as it was with some limited adaptation. Thus, the European Commission survey on climate change notes that individuals are tackling climate change with 70% trying to reduce waste and separate it for recycling, 55% trying to reduce their consumption of disposable items, 37% considering lower energy consumption whilst purchasing new household appliances, 31% buying and eating less meat, and 28% regularly using environmentally friendly alternatives to travel by private car (European Commission 2023, 35). All these are important adaptations, but collectively they make a minor contribution to alleviating the risks related to climate change.

4.3 Beyond the Risk Society Thesis

There is no question that Ulrich Beck's risk society thesis transformed understanding of the emergence of new forms of risk that emerge as society shifts toward a form of new modernity. Nevertheless, Beck's work fails to appreciate some of the features that are distinctive to new modernity that enhance risk and compound the impacts of industrialized threats. A key issue is to appreciate that no one risk, and any related damage, can be isolated from other risks and their direct, indirect, and induced effects (Bryson et al. 2024). Moreover, new forms of non-calculable uncertainty (Beck 1992) have emerged that result from the interplay of different and often unrelated but still linked socio-economic processes. This is to highlight that risks are increasingly manufactured and result from human interventions (Giddens 1999). There is a perversity to this process as one technological solution to some form of societal challenge, or commercial opportunity, may unleash new forms of risk. Central to this process is an alteration in the nature of risk that unfolds as a direct result of a major escalation in the complexity of the infrastructure that supports everyday living.

The risk society thesis is very much a social science debate constructed around a specific reading of risk and risk cultures. Another approach to understanding risk emerged in the work of Charles Perrow who was a professor of sociology at Yale University. Perrow is best known for his work on complex organizations and normal accidents and their relation to living with high-risk technologies (Perrow 1984, 1986). Perrow's analysis of complex organizations has been overlooked by geographers (but see Bryson et al. 2021), and yet his approach provides a conceptual explanation for many of the "risks as extra-network effects" that emerge from the configuration of global production networks (GPN) (Bryson and Vanchan 2020, 531). Extra-network effects are new risks that come from enhanced globalization, and these include rapid virus transmission that comes from increased global connectivity. To Perrow, new forms of risk emerge that are related to the ever-increasing complexity of the tightly coupled systems that support everyday living (Perrow 1984). The emphasis that he places on "tightly coupled systems" is critical and is a major addition to understanding new forms of industrialized or manufactured risk. Thus, a GPN is a tightly coupled system and contains all the new forms of risk that are identified by Perrow in his analysis of risk and complex organizations. It is unfortunate that these extra-network effects have been ignored by GPN researchers until comparatively recently (Bryson and Vanchan 2020; Völlers 2023; Völlers et al. 2023).

To Perrow (1984) these new forms of risk emerge based on two factors that enhance system susceptibility to risk. First, risk emerges in the multiple ways in which different parts of a system interact with one another. A system can be also conceptualized as a network. Some systems are linear meaning that any failure, or risk assessment, is obvious as direct and indirect effects can be forecast. A good example would be the failure of a bridge with consequent negative impacts on connectivity. Alternatively, much more complex systems have emerged that include multiple linked processes or systems. Risk assessment of these more complex systems is difficult and often impossible. These more complex systems are central to Beck's discussion of incalculable risks. The societal challenge with complex systems is that their parts may interact in unexpected ways resulting in the emergence of incalculable risks. To Bryson and Vanchan (2020) these risks include "extra-network effects" which reflect risks that lie beyond the network but are directly attributable to the configuration of a complex and interlinked network, for example, a GPN. The difficulty is that with complex systems much that occurs is hidden and incalculable. There are many examples of complex systems in which managers, owners or regulators no longer understand how the system operates or functions. One problem is legacy code embedded in highly complex computer programs involving many nested interrelationships. Legacy code is old computer source code that is no longer supported by current hardware and reflects a form of obsolete code. Legacy code is embedded in complex computer programs that support critical societal processes. A software update to a program with legacy code within it may result in unexpected failure. One of the challenges is that reading code produced by someone else is difficult as each coder will have idiosyncrasies and often coders fail to provide appropriate documentation (Thompson 2019). To computer programmers code that is "efficient and well organized is 'clean', 'beautiful', 'elegant' ..." whilst code that is a "slow, wretchedly designed mess" comes with a "bad smell" (Thompson 2019, 138). Code that stinks may work for a time, but contains many hidden risks (Fowler et al. 1999).

Second, Perrow (1984) highlights that risks related to complex systems are directly related to how much slack exists in a system. This concept of slack developed within the engineering literature on tight coupling or system optimization (Bryson et al. 2021). In this account of engineering systems, tightly coupled systems are those with limited slack, or buffering, existing between different parts of the system. Tightly coupled systems have been designed based around approaches to system optimization to enhance the cost-effectiveness of the engineered solution. The opposite of a tightly coupled system is one that is loose coupled in which slack exists in the system with the outcome being that failure in one part of a system might be covered by slack that exists elsewhere.

This concept of organizational slack is critical. There is a tendency for economic geographers to focus far too much on optimization whilst overlooking the relationship between tightly coupled value-generating processes, or networks, and risk. An excellent example of this approach to optimization in economic geography is found in Yeung and Coe's paper in which they seek to develop a dynamic theory of GPN, or GPN 2.0. In this journal paper, "a firm can therefore be thought of as a managerial device to optimize the accumulation and deployment of its available resources, defined as its core capability, at the lowest possible cost" (2015, 35). The use of the word "optimize" in this definition highlights the over-emphasize placed in the GPN literature on optimization without due acknowledgement of the risks that result from overtly tightly coupled networks (Vanchan, Mulhall, and Bryson 2018). One could argue that the Yeung and Coe definition is an accountancy-based definition that fails to acknowledge the critical role placed by organizational and network slack in reducing risks and enhancing corporate and network resilience. It is critical that economic geographers appreciate the links between risk and tightly coupled systems.

Organizational slack has been part of social science debates for over 60 years and yet economic geographers have tended to ignore this concept. The concept of organizational slack is central to the theory of the firm and was first identified in 1963 in Cyert and March's now classic work on organizational theory (Cyert and March 2001). This theory defined the contribution organizational slack makes to organizations by revealing that "when the environment becomes less favourable, organisational slack represents a cushion" and "absorbs a substantial share of the potential variability in the firm's environment" (Cyert and March 2001, 43). Organizational slack functions to stabilize organizations experiencing the outcome of some form of risk event. Organizational slack is too often defined as waste given the over-emphasis that is often placed on efficiency, value for money, performance, system optimization and productivity. Nevertheless, identifying and removing organizational slack results in the configuration of tightly coupled systems with the outcome being enhanced risk. There is a paradox here in that system optimization is a rational process, but it is also one that results in organized irresponsibility. It is important to appreciate that treating optimization as a core capability ultimately may result in systematic failure with the outcome being unexpected and incalculable costs. The concept of slack can be linked to debates on resilience which explore the ability of some entity (organizations, individuals, households, places) to resist or recover from some trauma. Resilience is the opposite of vulnerability defined as the degree to which an entity may react adversely to trauma or shock (Proag 2014). Slack provides an important source of buffering to potential or actual shock.

4.4 The New Jenga Capitalism Epoch

It is timely for economic geographers to develop a much more sophisticated approach to risk. There have been some recent developments that were prompted by the unexpected impacts of the COVID-19 pandemic on organizations and GPNs (see Bryson and Vanchan 2020). Nevertheless, now is the time to revisit and extend Beck's risk society thesis by developing a dialogue with Perrow's conceptual and empirical work on risks and tightly coupled systems. One of the challenges faced by all living in developed-market economies revolves around their on-going dependency on tightly coupled systems and this dependency is spreading to all countries. Writing this chapter requires access to an infrastructure system that generates and distributes electricity. However, this is only one of many infrastructure systems that are required to produce this chapter. The UK's National Grid, or the high-voltage electricity transmission network, is underpinned by the Web and computer programs. This means that any failure of the Web may impact on the National Grid and vice versa as the National Grid and the Web are coupled systems. The increasingly coupled nature of infrastructure systems enhances risk and represents a major escalation in Beck's "social explosiveness of hazards".

The extent of this escalation in the risk society highlights that a new period of capitalism has emerged that is best defined as Jenga Capitalism (Bryson et al. 2021; Bryson 2022). This new epoch of Jenga Capitalism reflects a very different form of modernity in which there is a significant uplift in "organized irresponsibility". In fact, organized irresponsibility transforms to a new form of disorganized irresponsibility. Tight coupling within national economies and GPNs is the outcome of an on-going cumulative process of system or network convergence. Added to this is that enhanced systemic connectivity based on a complex network of flows has emerged with the alignment of globalization with digitalization, webification and internetification. The outcome is the emergence of an increasingly and frighteningly tightly coupled system in which any disruption to part of this network of networks can lead to systemic disruption or collapse.

This new term – *Jenga Capitalism* – represents a new metaphor, but this is an action-based metaphor as all who have played Jenga are able to reflect on their experiences of playing this game and apply this to conceptualizing the stability of increasingly unstable socio-economic structures. Four processes define this new epoch of Jenga Capitalism. First, is the *incalculability of tightly coupled systems*, and their related risks and potential effects, that are over-optimized and have limited slack. These tightly coupled systems include all infrastructure networks – water, energy, information communications – and GPNs. Tightly coupled systems benefit from economies of scale which reduce risk but paradoxically also enhance societal exposure to new forms of risk. These new forms of risk include events that can be predicated with avoidance strategies developed and implemented combined with unimaginable risks that no one would ever assume could result in negative outcomes. With Jenga Capitalism the balance between these risks shifts to those that are unimaginable and the outcome being that very limited investment is made in mitigation and adaptation strategies. Examples of unimaginable risks include the impacts of a post COVID-19 global pandemic that would kill 60% of the human population, the complete failure of the Internet and/or global communication systems for a period of more than a month, or the commencement of World War III with nuclear strikes that destroyed over 150 world city-regions. One could argue that these events are not unimaginable, but the point being that companies, countries and households are not prepared for events of this geographic reach and intensity. Companies, governments, and households can prepare, but the question revolves around risk versus reward and the ability to take these types of risks and uncertainties seriously. It is important to appreciate that this exposure to all types of new risks and uncertainties applies to everyone and not just firms.

Second, is the on-going and dramatic escalation in *system convergence* based on planned or unplanned system convergence with the outcome being that new forms of unknowable risks have emerged. Central to this process are domino effects, or chain reactions, in which one part of a converged system of systems impacts on other parts. The point here is that with tightly coupled systems, systemic failure may emerge from some minor and unexpected fault in some aspect of the system. One example of this was the temporary blockage of the Suez Canal in March 2021 that disrupted GPNs for months (Bryson 2022, 221), and from 19 November 2023 Houthis, an Iranian-based rebel group, attacked commercial shipping in the Red Sea with drones, missiles, and speed boats further disrupting global supply chains. Another example is that within production systems, industrial control systems (ICS) have been developed that couple hardware with software through network interconnectivity. Network interconnectivity represents a weak link in production operational systems that can be exploited through cyberattacks (Bryson et al. 2022, 21). Historically, "production equipment was isolated or air-gapped from the outside world" (Bryson et al. 2022, 21). ICS are now optimized by integrating hardware with software with network connectivity. The outcome has been the emergence of a *cyber-energy-production plexus* (C-EPP) with the outcome being that cyberattacks have become one of the primary risks facing all ICSs that are coupled with the internet. This process of ICS internet coupling applies to all infrastructure systems that support everyday living.

The third point is that central to Jenga Capitalism is the central role played in everyday living, and in all production networks, of a *C-EPP* (Bryson et al. 2022, 21). For too long the social sciences have been over-fixated with networking thinking. The internet and energy networks have become a plexus; network thinking must be replaced with plexus thinking. A plexus is a structure consisting of a network of networks that are closely interwoven, intercommunicating, and coupled; any alteration within a plexus has the potential to ricochet throughout the structure producing unexpected consequences. The on-going formation of the C-EPP is a critical feature of Jenga Capitalism that exposes all individuals, organizations, and governments to new forms of risks and their possible impacts. This plexus forms around the configuration of multiple and extremely tightly coupled connections between telecommunications, energy, and production networks (Bryson et al. 2021). The IoT is central to the on-going deepening of the C-EPP as everything from cars to refrigerators and showers become coupled with the Web. Smartphones are one feature of this plexus with everyday living increasingly dependent on being linked to the C-EPP via a smartphone interface. Oliver Letwin a former UK cabinet minister, member of the UK's National Security Council and former minister with responsibility for the UK's national resilience has published an analysis of what could happen if one element of what I have termed the C-EPP failed.

His analysis is based on a UK scenario in which there is complete failure of the Web with the outcome being the death of hundreds of thousands of people. This scenario is possible, but limited political and media attention is being given to the risks related to increased societal dependence on the C-EPP. Thus, in the UK from December 2025 all households will have been switched from analogue phone landlines to an internet-based service. This is an excellent example of the on-going deepening of the C-EPP; analogue phone lines function during a power outage, but internet-based services are inoperable. The emergence of the C-EPP comes with many advantages, but the primary disadvantage is that the C-EPP is part of a complex ever-deepening plexus or system of interconnecting systems.

Fourth, geography or place and space play a central role in the new risks that emerge with Jenga Capitalism. There are many ways of defining globalization but perhaps the best definition is the on-going configuration of a *global plexus (GP)*; the *C-EPP* is a critical national subsystem of the GP. The GP is not to be confused with a GPN, or global value chain (GVC), as these are supply chains while the GP emerges from linkages that develop between more nationally configured but now globally linked C-EPPs and this includes all GPNs/GVCs. The GP operates at a very different scale and intensity compared to a GPN or GVC as the GP includes all devices and systems that are connected to the internet, and this includes devices or artifacts that are within an intranet but are still accessible via the internet. There is now no doubt that the developing GP is the most critical system of systems that enhances uncertainty and exposes all living on this planet to a new level risk. It is this GP that is central to the on-going intensification of risks and uncertainty that are a central feature of Jenga Capitalism. There are two core threats. On the one hand, a problem in one part of the GP has the potential to negatively impact on the functionality of connected, but unconnected systems. The critical importance of this GP is highlighted by cyber-attacks and the emergence of cyber warfare that highlights the disadvantages of nationally configured C-EPPs being incorporated into the GP. Letwin, for example, highlights the impacts that cyberattacks had on Ukraine in 2016 and 2017. On 23 December 2015, a preliminary cyberattack on Ukraine's infrastructure resulted in 230,000 people being without electricity (Letwin 2020, 33). This was followed by an attack in December 2016 that took out part of Ukraine's state energy computer system. Then, on 27 June 2017 a much more sophisticated attack "closed down Ukrainian ministries, banks, airports, railways, metro systems and telecom services – a whole range of key networks" (Letwin 2020, 33). Cyberattacks on a nation's internet impact on other networks given the system convergence that is central to the on-going development of each nation's *C-EPP*, but these national plexuses are then coupled with the GP. It is critical that social scientists explore the on-going configuration of the GP to unravel potential weaknesses and to identify risk avoidance strategies.

This ever-deepening and expansion of the GP enhances every country's exposure to enhanced and uncontrollable risks and this applies to all individuals and companies whose everyday actions depend on this evolving plexus. A central characteristic of Jenga Capitalism is the geographic reach of the risks combined with the duration of events that result in societal damage. During this new epoch of Jenga Capitalism, a problem that emerges in one locality can be rapidly transmitted to other places. What in earlier times would have remained a local problem may in the twinkling of an eye become a global problem. This highlights the need to develop a multi-scalar approach to understanding the transformation of Beck's risk society into Jenga Capitalism. It is important to appreciate the complexity of place as each place has a distinct accumulation of resources and each place is connected to other places in very different ways. The implication being that some places are more interconnected with the GP and other places are less connected. Nevertheless, even less connected places are exposed to Jenga Capitalism risks.

Jenga Capitalism enhances the precarity of everyday living and of production systems. Surviving within Jenga Capitalism requires the application of three strategies. The first is to ensure that alternative infrastructure systems, and control mechanisms, are in place that would replace infrastructure service provision when it is compromised due to the failure of some element of the C-EPP. Second, some critical control mechanisms must be isolated or air-gapped from the C-EPP to protect them from cyberattacks and this air-gapping must be applied to every country's C-EPP to remove some of the risks related to over-coupling with the GP. Alternatively, computer-facilitated systems must be designed to ensure that they can be operated manually. Third, nations must develop and implement initiatives intended to enhance their national sovereignty. This includes ensuring that critical products and services are produced locally as well as being sourced globally. It is important to ensure that no one country becomes too dependent on another for any good or service that is critical for national security and for ensuring the continuation of everyday living. There are many examples of government strategies intended to enhance national economic sovereignty. For example, President Jinping Xi, China's paramount leader since 2012, emphasizes the strategic importance of China becoming self-reliant in food and he once stated that "Chinese people should hold their rice bowls firmly in their own hands, with grains mainly produced by themselves" (Xi as quoted in Manickam 2022). Similarly, both the European Union and the US have developed and implemented initiatives based on digital sovereignty to ensure that their citizens are not overdependent on semiconductor chip-making plants located in other national jurisdictions (Bryson 2022).

4.5 Theorizing Risk and Economic Geography

The risk society is in a continual state of transition as new forms of risk emerge that are part of the shift into the Jenga Capitalism epoch. The conceptual and empirical social science literature needs to develop a much more sophisticated approach to understanding and researching risk. This includes exploring the ways in which individuals, households, organizations, and regional and national governments engage in activities intended to enhance certainty and reduce exposure to the possible negative impacts of risks. Much of the economic geography literature, for example, has "emphasised intra-network analysis rather than exploring the much wider inter- and extra-network risks associated with globalisation" (Bryson and Vanchan 2020, 535). Globalization, or the on-going configuration of the GP as an assemblage of coupled C-EPPs, creates new forms of socio-economic risks, or extra-network risks, and these risks remain unaccounted in much of the economic geography literature.

In economic geography, the GPN framework has become too dominant with this being the conventional approach applied to understanding the economic geography of globalization. Globalization is a very complex process that has multiple positive and negative impacts. The GPN approach tends to celebrate the positives and ignore the negatives (see Blažek 2016; Bryson and Vanchan 2020; Werner 2019). Central to this approach is the assumption that "there is now no doubt that global production networks (GPN) and global value chains (GVC) are the most critical organizational platforms through which production in primary, manufacturing, and service sectors is coordinated and organized on a global basis" (Yeung and Coe 2015, 30). This is rather an extreme statement as it discourages alternative approaches to be identified and explored, for example, online platforms or localized approaches to configuring production. GPNs and GVCs are responsible for encouraging over-consumption, emissions of greenhouse gases that contribute to climate change, linear production systems that create wasteful forms of consumption and for their involvement in labor exploitation including modern-day slavery. There is another problem with this statement in that it assumes that companies apply GPN/GVC thinking in practice. The reality is that companies will apply some of the more technical and quantitative approaches to supply chain management that have been developed in the operations literature (Bryson and Vanchan 2020). It is important that economic geography does not overstate the role played by GPNs or GVCs and does apply a critical approach that does demonstrate the ways in which these academic debates make measurable impacts that reduce some of the negatives that are the outcomes of GPNs and GVCs. This includes avoiding too much of an emphasis on the internal workings of GPNs and assuming that the benefits of this approach to organizing production outweigh the negatives. What is required is a "more balanced geography of complex production systems" that acknowledges "the heterogeneity of production forms and systems" including the contributions made by microfirms as well as larger corporate forms (Rusten and Bryson 2010, 258).

It is important that economic geography encourages critical debate on globalization, and this includes developing alternative approaches to exploring production processes in place and across space (Blažek 2016). Recently, such an alternative approach has been developed. This is the *TRAcking Corporations Across Space and Time* (TRACAST) framework that provides an approach to systematically identifying and exploring companies within supply chains and this includes identifying environmental and societal impacts as well as key nodes of governance. The starting point for this new framework is the realization that:

> we often lack knowledge of how our consumption affects the environmental and socioeconomic conditions of distant peoples and places. Consider a global supply chain that produces the batteries in electronics sold by Apple, Dell, LG, and Samsung (Amnesty International, 2016). Cobalt mines in the Democratic Republic of Congo provide the raw material for a complex network of trading houses and smelters. These mines employ children in dangerous jobs and degrade the natural landscape. Battery component manufacturers in Asia purchase the refined cobalt, which ends up in computers sold by these transnational corporations.
>
> (Goldstein and Newell 2020, 1)

This example includes societal risks that are realized through negative societal impacts which in this case include child labor and environmental impacts. It is important that economic geographers appreciate the relationship between risks that remain possibilities and risks that result in direct and indirect negative impacts that can be measured and evaluated. The starting point for the development of the TRACAST approach was a recent review of the global commodity chains and production networks literature that revealed that "less than 1% of studies (27 out of ~11,000)" had explored "the supply chains of individual corporations" (Goldstein and Newell 2020, 2). Research had explored individual companies, but the focus had not been on their supply chain linkages. This omission "hinders our ability to better understand globalization processes and their impacts" (Goldstein and Newell 2020, 2). TRACAST provides a new approach to identifying and measuring these impacts and it is also an approach that is intended to "assist corporate actors committed to transparency and shifting toward more environmentally and socially responsible sourcing" (Goldstein and Newell 2020, 213).

The development of GPN 2.0 comes with an appreciation of intra-network risks and Yeung and Coe argue that "these risk forms require further theorization" (2015, 41) and that "the qualitative nature and causal effects of risk play out differently in the context of global production networks such that we can term it GPN risk" (2015, 41). There are many difficulties with this statement including the failure to acknowledge that GPN configurations creates new forms of negative societal, environmental, and political risk. Political risk emerges when one country or firm becomes over-dependent on components, resources and capacities located in another national jurisdiction. There is a fundamental problem with the ways in which GPN 2.0 conceptualizes risk as the assumption is that "risk is generally produced beyond the control or confines of individual actors" (Yeung and Coe 2015, 41). This statement only applies to some types of risk, or a sub-set of risks, as there are risks that are firm and even product- or service-line specific and which form through the actions or activities of individual actors. One of the challenges is that economic geography has failed to develop a dialogue with the operations management literature (Bryson 2022; Vanchan, Mulhall, and Bryson 2018). This literature is much less focused on theorizing, but instead develops tools and frameworks that are applied in everyday practice.

All well-managed companies apply risk management processes, and all should have a risk assessment matrix in place. The development and implementation of this organizational approach to risk is a critical micro-foundational process that has been too overlooked by economic geographers (Bryson and Vanchan 2020; Fuller 2023; Völlers 2023; Völlers et al. 2023). This is a micro-foundational process as a risk matrix is the outcome of a decision-making process based on people assessing all types of risks. Such a matrix identifies a multitude of risks with the matrix being regularly reviewed and updated. These "risk grids are context-dependent" and will include "key risks, likelihood, impact, numeric score, controls, responsibility, further action" (Bryson 2022, 223). Each risk is reviewed including mitigation and adaptation strategies and an individual or group within the firm is named as having responsibility for overseeing each risk. These risk matrices include risks that are completely under the control of the firm and those that depend on the firm negotiating with other actors. Nevertheless, well-managed firms will have identified actions that are within their control that might alleviate these risks. In any case, there are many examples of interesting approaches to the management of risk, for example the insurance industry applies a risk equation (see Omand 2023, 21):

$$\text{Level of risk} = \text{likelihood} \times \text{vulnerability} \times \text{initial impact} \times \text{duration of disruption}$$

This is a very technical process with what is known in the insurance industry as the product of the risk equation revealing high-risk levels requiring investment in mitigation measures. With the emergence of Jenga Capitalism, there are uncertainties that cannot yet be imagined as risks and thus no risk equation can be produced. Understanding how risk equations are calculated, and their impacts, should be a critical area for future socio-economic and geopolitical research.

Risks are sector and company-specific, but there are risks related to dependency (key employees, other firms), recruiting and retaining employees, skill deficiencies including those at the level of the board, complying with statutory regulations, external environment, sales/revenues, health and safety (employees, customers and visitors), cybersecurity, failure of core systems, disruptive innovation, legal action, industrial action, inappropriate use of company resources, quality issues, financial control, emergence of unanticipated and uninsured costs, catastrophic loss of facilities, server failure, competition, government policy, reputational damage, social media and public perception. Each firm will be exposed to risks that are outside their direct control, but each firm will have a different degree of exposure depending on decisions that they have made in the past combined with their current approach to ameliorating risk (Bryson 2022; Bryson and Ronayne 2014; Vanchan, Mulhall, and Bryson 2018).

There is a major conceptual lacuna in existing approaches to exploring risk in economic geography. Risks are too often treated in isolation. At any one time a firm, region or country is exposed to multiple risks and some of these are actualized into impacts. It is also important not to overlook induced impacts related to investment intended to ameliorate risk that prevents investment in other areas. Economic geographers working on companies and/or regions/globalization must explore the operation of "shock chains and parallel shocks" (Bryson et al. 2024). Shock chains, or concatenations of shocks, refer to shocks that occur in sequence and may be related or unrelated to one another. One outcome of the existence of shock chains is that there are also recovery chains, or recovery processes that occur in sequence, but with overlapping stages. Shock chains emerge from risks whose impacts have not been avoided and reflect a combination of exogenous and endogenous shocks. An excellent example is the pandemic whose impacts continue to disrupt, but such disruptions are then further complicated by new shocks that might be directly or indirectly linked in some way to the pandemic or these new shocks may be completely unrelated to the pandemic.

Shocks of different intensity also occur in parallel, and this is also the case for recovery or adaptation processes (Bryson et al. 2024). At any one time, a country, region, individual, group, firm or GPN will be experiencing multiple parallel shocks. Some shocks will have impacts that evaporate at the moment that the shock occurs, and there will be no lasting impacts requiring recovery as recovery is instantaneous. Other shocks alter circumstances permanently, for example, the death of a key manager or business leader, or the closure of a supplier, and recovery will then need to be based on adaptation to the new circumstance. Multiple feedback loops may also emerge with parallel shocks and further complications emerge from the impacts of shock chains.

No one firm, GPN or city-region experiences one risk or the impacts from one risk, but instead is exposed to a complex array of risks and these risks materialize in the form of shock chains or parallel shocks. Understanding and conceptualizing this complexity must become central to on-going debates on the economic geography of firms, regional economies, globalization, and labor. What is required is the development of a new conceptual framework for understanding firms in place that adopts an approach that includes a focus on macro-processes, but also one that acknowledges the importance of understanding the micro-foundations that underpin macroeconomic phenomena based on understanding individual and group behaviors and their interactions. It is important that economic geography places people at the center of its analysis: it is people who make assessments regarding risks that are then reflected in investment decisions and the decisions that underlie the configuration of all forms of production network.

4.6 Conclusion

Firms and regions are in a continual state of becoming as they adapt to endogenous and exogenous pressures, changes, risks, and shocks. Every company is the outcome of a plexus of intra- and inter-company processes that are framed within different contexts. All companies are positioned within wider structures and any one individual company process, or business activity system (marketing, design, operation, supply chain management, logistics, finance), is part of a much more complex array of interdependent corporate processes. Each process will be subjected to specific risks but framed with the context of some overarching company risks. All this suggests that the definition of a firm used by economic geographers must include an appreciation that all business activities come with risks as well as creating risks that are experienced by other companies, individuals, and places. Definitions matter and the definition of the firm must include an awareness of tight and loose coupling and the importance of organizational slack.

It is time for economic geography to treat risk seriously. This includes differentiating between risks and avoidance processes and risks that result in negative and positive impacts. A risk becomes an impact if some agent experiences an outcome that might come from an imagined or unimagined shock. Developing an approach to risk in economic geography requires understanding macro-level changes coupled with an appreciation of micro-foundational processes. At a macro-level, Beck's risk society thesis is a useful starting point for exploring new forms of risk. Nevertheless, this thesis dates back to 1986 and there have been significant societal alternations that have transformed capitalism with the outcome being the on-going configuration of Jenga Capitalism. Jenga Capitalism is partly an outcome of technological innovation and the failure of governments and innovators to acknowledge the escalation in risks that are related to deepening globalization that is the outcome of the system convergence that lies behind the formation of a GP and the configuration of more nationally orientated C-EPPs.

It is important to appreciate that all corporate activities come with risks. This is an important point. The Merchant of Prato's business activities were saturated with risks and sometimes these risks were actualized into impacts. Some of this merchant's risks are also experienced by companies operating in 2024, but there are qualitative and quantitative differences. It is important to appreciate the complexity of risks, shocks, and impacts. New forms of risk have emerged related to tightly coupled systems and that are also the outcome of optimization approaches applied by individual companies as well as governments. It is important to recognize that all firms, regions, and governments are exposed to a concatenation of risks which are experienced in parallel and in sequence or chains. Thus, it is important to explore and conceptualize the role played by shock chains and parallel shocks. The key here is to acknowledge risk complexity. There is a danger that economic geographers become too distracted by one shock, for example, a financial crisis, flood or pandemic.

This danger is also experienced by companies and governments who suffer from the distractions of the immediate problem or risk that is currently creating major societal problems or challenges. The danger is that over-focusing on the distraction of the immediate challenge, for example, the COVID-19 pandemic, may prevent initiatives and investments required to mitigate other risks.

The outcome of this analysis of the changing economic geography of companies and regions in times of risk, uncertainty, and crisis is a call for a more balanced approach to develop in economic geography (Rusten and Bryson 2010). This must be a multi-scalar approach and one that acknowledges the interplay between macro processes and the micro-foundational processes that underpin all economic activities. This is a call for a new approach to theory building that acknowledges the diversity of company forms and the variegated nature of regions. It is also a call for social scientists to move beyond simple network thinking to embrace the complexity that comes with understanding plexuses. It is also a call for theory-led empirical research that will explore risk, uncertainty, and crisis at different spatial scales and in different sectors and geographic contexts. This research should explore shocks that lead to transformational innovation based on the assumption that the negative impacts of some shocks encourage potentially beneficial disruptive innovations.

References

Beck, U. 1992. *Risk Society: Towards a New Modernity*. London: Sage.

Beck, U. 2000. Risk Society Revisited: Theory, Politics and Research Programmes. In *Risk Society and Beyond: Critical Issues for Social Theory*, ed. B. Adam, U. Beck, and J. Van Loon, 211–29. London: Sage.

Blažek, J. 2016. Towards a typology of repositioning strategies of GVC/GPN suppliers: the case of functional upgrading and downgrading. *Journal of Economic Geography* 16 (4):849–86. doi: 10.1093/jeg/lbv044

Bryson, J. R. 2022. Reading Manufacturing Firms: Lost Geographies of Value/Risk and the emergence of Jenga Capitalism. In *A Research Agenda for Manufacturing Industries in the Global Economy*, ed. J. R. Bryson, C. Billing, W. Graves, and G. Yeung, 211–44. Cheltenham: Edward Elgar.

Bryson, J. R., L. Andres, A. Ersoy, and L. Reardon. 2021. A year into the pandemic: shifts, improvisations and impacts for people, place, and policy. In *Living with Pandemics: Places, People and Policy*, ed. J. R. Bryson, L. Andres, A. Ersoy, and L. Reardon, 2–34. Cheltenham: Edward Elgar.

Bryson, J. R., L. Andres, A. Ersoy, and L. Reardon. 2024. Shock Chains and Parallel Shocks: Towards a Social Science of the Recovery Society. In *Pandemic Recovery: Reframing and Rescaling Societal Challenges*, ed. J. R. Bryson, L. Andres, A. Ersoy, and L. Reardon, 1–26. Cheltenham: Edward Elgar.

Bryson, J. R., C. Billing, W. Graves, and G. Yeung. 2022. Reframing manufacturing research: place, production, risk and society. In *A Research Agenda for Manufacturing Industries in the Global Economy*, ed. J. R. Bryson, C. Billing, W. Graves, and G. Yeung, 1–32. Cheltenham: Edward Elgar.

Bryson, J. R., and M. Ronayne. 2014. Manufacturing carpets and technical textiles: routines, resources, capabilities, adaptation, innovation and the evolution of the British textile industry. *Cambridge Journal of Regions, Economy and Society* 7 (3):471–88. doi: 10.1093/cjres/rsu018

Bryson, J. R., and V. Vanchan. 2020. COVID-19 and Alternative Conceptualisations of Value and Risk in GPN Research. *Tijdschrift voor Economische en Sociale Geografie* 111:530–42. doi: 10.1111/tesg.12425

Cyert, R. M., and J. G. March. 2001. *A Behavioural Theory of the Firm*. Oxford: Blackwell.

European Union. 2023. *Special Eurobarometer 538 Climate Change*. Brussels: European Commission.

Fowler, M., K. Beck, J. Brant, W. Opdyke, and D. Roberts. 1999. *Refactoring: Improving the design of existing code*. Boston: Addison-Wesley.

Fuller, C. 2023. Uncertainty, fictional expectations and economic agency. *Geoforum* 140:103699. doi: 10.1016/j.geoforum.2023.103699

Giddens, A. 1999. Risk and Responsibility. *Modern Law Review* 62 (1): 1–10.

Goldstein, B., and J. P. Newell. 2020. How to track corporations across space and time. *Ecological Economics* 169:106492. doi: 10.1016/j.ecolecon.2019.106492

Knight, F. H. 1921. *Risk, Uncertainty and Profit*. New York: Sentry Press.

Letwin, O. 2020. *Apocalypse How? Technology and the Threat of Disaster*. London: Atlantic Books.

Manickam, G. 2022. My account on China: China's quest to secure its 'rice bowl'. *CGTN*, July 26. Accessed July 22, 2023. https://news.cgtn.com/news/2022-07-26/My-account-on-China-China-s-quest-to-secure-its-rice-bowl--1bXGBcHvvRS/index.html

Omand, D. 2023. *How to survive a crisis*. London: Viking.

Origo, I. 1988. *The Merchant of Prato*. London: Penguin.

Perrow, C. 1984. *Normal Accidents: Living with High Risk Technologies*. Princeton, NJ: Princeton University Press.

Perrow, C. 1986. *Complex Organizations: A Critical Essay*. New York: McGraw-Hill Publishers.

Proag, V. 2014. The Concept of Vulnerability and Resilience. *Procedia Economics and Finance* 18:369–376. doi: 10.1016/S2212-5671(14)00952-6

Rusten, G., and J. R. Bryson. 2010. Placing and Spacing Services: Towards a balanced economic geography of firms, clusters, social networks, contracts and the geographies of enterprise. *Tijdschrift voor Economische en Sociale Geografie* 101:248–261. doi: 10.1111/j.1467-9663.2009.00584.x

Thompson. C. 2019. *Coders*. London: Picador.

Vanchan, V., R. Mulhall, and J. R. Bryson. 2018. Repatriation or Reshoring of Manufacturing to the U.S. and UK: Dynamics and Global Production Networks or from Here to There and Back Again. *Growth and Change* 49: 97–121. doi: 10.1111/grow.12224

Völlers, P. 2023. Mitigating Extra-Firm Risk Environments – The Case of Turkish Firms in Germany. *Tijdschrift voor Economische en Sociale Geografie* 114:237–251. doi: 10.1111/tesg.12557

Völlers, P., T. Neise, P. Verfürth, M. Franz, F. Bücken, and K. P. Schumacher. 2023. Revisiting risk in the Global Production Network approach 2.0 - Towards a performative risk narrative perspective. *Environment and Planning A: Economy and Space* 55 (8):1838–58. doi: 10.1177/0308518X231169288

Werner, M. 2019. Geographies of production I: Global production and uneven development. *Progress in Human Geography* 43 (5):948–958.

Yeung, H., and N. Coe. 2015. Toward a dynamic theory of global production networks. *Economic Geography* 9 (1):29–58. doi: 10.1111/ecge.12063

5 Regional Economic Resilience, Uncertainty, and Fictional Expectations

Pieter Terhorst and Hilal Erkuş

5.1 Introduction

In capitalism regions and national states go through periods of crisis and stagnation from time to time. Shocks are not only the result of inherent capitalist dynamics but also of geopolitical and environmental crises such as wars, earthquakes, and pandemics like COVID-19. Regional economic resilience literature focuses on how some countries and regions are more resistant to shocks and are more successful in recovering from them than are others. Some regions are less shock proof and bounce back from a shock whereas other regions bounce forward from it (Martin 2012, 2018; Martin and Sunley 2015, 2020).

Much regional economic resilience literature, dominated by evolutionary economic geography, conceived regional economies as complex adaptive systems, in which the role of agency is under-explored (Bristow and Healy 2014a; Pike, Dawley, and Tomaney 2010). Over the last years, however, several authors call for more attention to agency in regional economic resilience, and argue that it is driven by a combination of structural and agentic factors (Bristow and Healy 2014a, 2020; Evenhuis 2017; Grillitsch and Asheim 2023; Magro, Uyarra, and Valdaliso 2022; Tan et al. 2020).

This chapter explores regional economic resilience mainly from the perspective of one dimension of agency. It aims to connect the concept of regional economic resilience with the concept of fictional expectations about the future, a concept coined by the leading economic sociologist Beckert (2016). Fictional expectations refer to "the images actors form as they consider future states of the world, the way they visualize causal relations, and the ways they perceive their actions in influencing outcomes" (Beckert 2016, 9). Beckert (2016) and Beckert and Bronk (2018) are not the only ones who point to the important role expectations, imaginaries, and narratives play in economic life. Economists Akerlof and Shiller (2009), and Köhn (2017, 177-192) and, more recently, economic geographers Fuller (2023) and Völlers et al. (2023) do.

Both in "normal" times and crises entrepreneurs and policy makers have to form expectations about the future and act accordingly. In "normal" times they have to anticipate to possible shocks in the future, and when a region is hit by a shock, they have to interpret the cause of a shock, which is open to

DOI: 10.4324/9781003457626-6

interpretation, debate, and political struggle. Expectations about the future are, to a large extent, shaped by the dominant interpretation of the cause of a shock. Do they expect a return to the pre-shock development and can therefore act in the same way as before? Or do they interpret the shock as an indication that a new development path has to be chosen? In both cases, they have to form expectations about the future. In doing so, however, they face radical uncertainty in the sense of Knight (1921), i.e. they cannot ground their expectations on a probability calculus, because the future is open.

Economic and political decision-making is impossible without any belief in how the future will look like. Entrepreneurs have to form expectations about the returns on their investments, the results of their innovative strategies, the strategies of other entrepreneurs as well as the development of the world around them. The same applies to policy makers: they have to form expectations of the effects and efficiency of their policies and the future of society as a whole. Fictional expectations about the future, which are the antipode of neoclassical concept of rational expectations (Lucas 1972), provide some anchorage in decision-making under conditions of uncertainty and form a motivational force of action. They take the form of narratives and scenarios that steer the regional strategies to bounce back or bounce forward from shocks.

Fictional expectations are not only individually but also, and to a large extent, socially constructed by a variety of social factors that are both historically and geographically specific. Place-based characteristics of regional economies shape the fictional expectations of local actors but do not determine them, i.e. they shape the expectations of possible opportunities and constraints in the future. This is illustrated in our two case studies. The first one is about the anticipatory adaptation of the Port Authority of Rotterdam to geopolitical tensions and the rise of the sea level in the future, and the second one is about fictional expectations of post-COVID-19 tourism in Antalya, a sun-sand-sea mass-tourism city that was severely hit by the COVID-19 crisis.

The organization of this chapter is as follows. Regional economic shocks are the outcome of both the instabilities of capitalism and environmental and geopolitical shocks. They have always been there, but, as argued in Section 5.2, it is highly likely that the current phase of capitalism is more unstable than the former post-war, Fordist phase, and that environmental shocks generated by climate change and epidemics like COVID-19 have increased. These instabilities are the *prima causa* of regional economic shocks and radical uncertainty about the future among business (wo)men and policy makers. In Section 5.3, we make a few remarks on regional economic resilience ending with the remark that the call for agency has become louder recently. In Section 5.4, we make a short note on the relation between structure and agency, and present Emirbayer and Mische's (1998) three dimensions of agency. In Section 5.5, we discuss the controversy on uncertainty between mainstream and heterodox economists. It comes down whether or not a probability calculus is possible in economic life. In Section 5.6, we discuss the concept of fictional expectations, and in Section 5.7, we bring fictional expectations in regional economic resilience as shown in our cases on the Port Authority of Rotterdam and Antalya.

5.2 Economic, Geopolitical, and Environmental Instabilities

Capitalist economies are very dynamic. They are in the words of Schumpeter ([1942] 1976, 81–86) in a continuous process of "creative destruction". New goods and services, new markets, new production methods, new forms of capital-labor relations, inter-firm relations, new types of firm organization, and economic institutions displace old ones, and it is these processes of creative destruction that make capitalism dynamic. Capitalism has always been dynamic but there are good reasons to believe that in the current form of globalized, financialized, cultural-cognitive form of capitalism instability is larger than ever. The deepening of the global division of labor has created strong structural interdependencies between locations all over the world, so that the spatial scale of the impact of shocks in one part of the world has become much larger than before. In addition, the financialization of the economy implies that surplus capital has increasingly been channeled toward speculative, casino-like pursuits, rather than long-term investments in the real economy; and it has made many financial products so complex that even most insiders in the financial world hardly understand them anymore as the global financial crisis of 2008 has shown. Finally, in the current phase of cultural-cognitive form of capitalism (Scott 2012) the share of markets of symbolic goods and so-called experience goods, i.e. goods the quality of which can only be known after consumption (holiday trips, restaurants, wines, theater- and concert performances, and expositions among many others) has increased. Their quality stems largely from their symbolic qualities based on interpretation and belief rather than from their objective characteristics. The fact that their standards of quality are often contested, accompanied with continuously changing fashions and fads, makes the markets of those goods highly volatile.

These economic instabilities have been accompanied by climate change that seems to generate heavier storms, tornados, floods, and droughts than before, and by increased geopolitical tensions between the West on the one hand and China and Russia on the other. Thus, it seems we are facing simultaneously multiple crises that are interacting: a polycrisis (Tooze 2022).

5.3 Regional Resilience: Toward More Agency

Afore-mentioned instabilities generate regional economic shocks, the impact of which varies between regions. Original conceptions of resilience were based on engineering and ecological traditions, emphasizing the ability to withstand a shock or, failing this, to recover (or bounce back) from a shock. In the course of time, evolutionary economic geographers, inspired by evolutionary theories of economic change and models of complex adaptive systems, began to stress the importance of both the ability of actors to adapt to changing circumstances and for the economic system itself to transform. Leading evolutionary economic geographers Martin and Sunley (2020) distinguish two evolutionary economic conceptions of resilience. (1) Evolutionary or adaptive resilience which refers to the ability of an economy to maintain its core function by reorienting and reorganizing its economic structure to a new and more successful growth path. Thus, in this case, an economy successfully adapts its structure to a shock without losing its core function. It bounces forward when hit by a shock. (2) Transformative resilience is more radical and refers to the ability of an economy to reconfigure and create a new economic structure and function once environmental, social, or economic conditions make the current structure and function untenable or undesirable. Their distinction largely overlaps with the processes of adaptation and adaptability. The former is related to maintaining regional growth paths and the latter refers to the creation of new paths (Boschma 2015; Pike, Dawley, and Tomaney 2010).

Regional economic resilience has raised a lot of debate among economic geographers and has generated a voluminous amount of literature (Martin 2018; Martin and Sunley 2020; Evenhuis 2017; see Sutton et al. 2023 for an exhaustive and systematic overview). Here we limit ourselves to a few remarks that, given the aim of this chapter, are most important to us.

Martin (2018) argues that regional economic resilience should not be interpreted as a static property of a regional economy that is only present when a shock occurs. He stresses that it is a feature that is present in both shock and non-shock periods, be it that the processes of "creative destruction" in shock periods are stronger than in non-shock periods. Nevertheless, also in "normal" (non-shock) periods regional economies adapt themselves to changing circumstances and innovate. The higher the adaptability and innovativeness of a regional economy in "normal" times is, the less vulnerable it is to shocks and the more successfully it recovers from shocks. In this way, low vulnerability to and high recoverability from shocks reinforce uneven regional development.

Early literature on regional economic resilience was strongly firm-centric and has been inspired by the theory of complex adaptive systems (Sutton and Arku 2022 advocate this approach) or, more precisely, by a one-sided interpretation of it. Complex adaptive systems are composed of a collection of heterogeneous actors who interact and adapt to each other and to the environment to produce a behavior of the system that cannot be explained in terms of the behavior of individual actors. Systems have, in terms of critical realist philosophy, an emergent property, i.e. the powers and liabilities of the system cannot be reduced to those of their constituents. Systems are characterized by complex non-linear dynamics and an adaptive capacity that enables them to rearrange their internal structure spontaneously whether in response to an external shock or an internal crisis. Although this approach does not necessarily exclude agency, most researchers on regional economic resilience have been focused on the system's structure, performance, and overall functioning when hit by a shock, and explain variety of regional economic resilience on the basis of place-based characteristics created in the past, such as industrial and business structure, pre-regional economic growth path, regional position in global economy, government arrangements, state–business relations, labor market, national macro-economic conditions, and level of urbanization among others (Brakman, Garretsen, and Van Marrewijk 2015; Capello, Caragliuy, and Fratesi 2015; Crescenzi, Luca, and Milio 2016; Giannakis and Bruggeman 2017; Martin 2012; Martin and Sunley 2015).

Evolutionary economic geography has been criticized for being too firm-centric while ignoring political institutions and politics in regional economic development as well as other non-firm actors like labor unions, business associations, universities, research institutes, and community organizations at multiple scales. Several studies have pointed out the positive role of politics in regional economic resilience (Bristow and Healy 2014b; Cowell, 2013; Davies 2011; Eraydin 2016; Evans and Karecha 2014; Kakderi and Tasopoulou 2017).

Whereas structure, such as regional industrial composition and political-economic institutions, was prioritized over agency for some time, economic geographers have recently called for more attention to agency in regional economic resilience, just as we do in this chapter (Bristow and Healy 2020; Evenhuis 2017; Grillitsch and Asheim 2023; Magro, Uyarra, and Valdaliso 2022; Tan et al. 2020). The same applies to the analysis of regional development paths in the innovative work of Jolly, Grillitsch, and Hansen (2020), Grillitsch and Sotarauta (2020), and Sotarauta and Grillitsch (2023) (although closely related to regional economic resilience, a discussion of regional development paths falls beyond the scope of this chapter). Paying more attention to the agency does not imply that structure is ignored. Not at all, all those authors try to combine structural and agentic explanations.

5.4 A Short Note on Agency and Structure

The relation between agency and structure can be very shortly clarified with Marx' famous statement in his *18th Brumaire* that "men make their own history but they do not make it just as they please; they do not make it under circumstances chosen by themselves but under circumstances directly encountered, given and transmitted from the past" (Marx [1852] 1995, 5). We are born into sets of relationships such as relations of production, inherited institutions, distribution of resources (capital), ways of thinking, perceiving, expecting, and acting which pre-exist with us. Structure may enable and constrain agency but there is no structure without agency, both past and present. Actions shaped by the past do not preclude agency because the past influences and shape actions without wholly determining them. It is tempting to interpret structure as a product of a static past but this is misleading. Actions lead to intended and unintended consequences that become part of a new past that offers new constraints and opportunities, and so on. Thus, history is a continuous process with no finished "products". Structure is always in process (see for more Bashkar 1979, 51–77).

Temporality also plays a key role in Emirbayer and Mische's (1998) seminal paper on agency. Agency is a process that happens in the presence and is shaped by historically developed structures and perceived futures. They conceptualize agency as "a temporally embedded process of social engagement, informed by the past (in its 'iterational' or habitual aspect) but also oriented toward the future (as a 'projective' capacity to imagine alternative possibilities) and toward the present (as a 'practical-evaluative' capacity to contextualize past habits and future projects within the contingencies of the moment)" (Emirbayer and Mische 1998, 962). Thus, they distinguish three dimensions of agency, namely (i) iteration, (ii) projectivity, and (iii) practical evaluation.

By (i) *iteration* they mean "the selective reactivation by actors of past patterns of thought and action, as routinely incorporated in practical activity, thereby giving stability and order to social universes and helping to sustain identities, interactions, and institutions over time" (Emirbayer and Mische 1998, 971). Thus, iteration refers to habitual and routinized reproduction of structures.

(ii) *Projectivity* is "the imaginative generation by actors of possible future trajectories of action, in which received structures of thought and action may be creatively reconfigured in relation to actors' hopes, fears, and desires for the future" (Emirbayer and Mische 1998, 971). Beckert's (2016) fictional expectations are an elaboration of Emirbayer and Mische's projectivity.

(iii) *Practical evaluation* entails "the capacity of actors to make practical and normative judgments among alternative possible trajectories of action, in response to the emerging demands, dilemmas, and ambiguities of presently evolving situations" (Emirbayer and Mische 1998, 971). In contrast to iteration, which largely rests on learned patterns from the past, and projectivity which rests on the imagination of the future, it is mainly situated in the present. It refers to conscious and critical deliberation of the situation actors find themselves in, often in dialogues with others. In times of a regional economic shock, actors have to interpret the cause of a shock in relation to the past and imagination of the future in order to adequately respond to it. Most of the literature of politics of regional economic resilience implicitly assumes that crises have a materialist and objective basis only, and ignore, therefore, that they simultaneously require a subjective and ideological interpretation. Was the cause of the crisis of the 1970s too high welfare state expenditures (neoliberals), the end of the efficacy of post-war basic innovations (neo-Schumpeterians) or a Fordist mode of regulation did not fit anymore to its accumulation regime (French regulationists)? And policy makers have to judge who pays the price for and who benefits from their resilience policy. Can we speak of regional resilience when the share of wages in the growing regional product remains stable or declines, and environmental degradation increases? "Resilience for whom?", Hudson (2010) wonders.

These three dimensions of agency do not exclude each other. Actors in the present are always simultaneously oriented toward the past and the future, although they may be primarily oriented toward one of them in specific situations. Beckert's (2016) fictional expectations are an elaboration of projectivity and are mainly situated in the present and the future. But, as we shall see below, he has an open eye to iteration, so that fictional expectations can be better linked to historically and geographically specific structures. What is missing in Emirbayer and Mische's (1998) analysis of projectivity, however, is a discussion of radical uncertainty actors face in the future.

5.5 Radical Uncertainty Highly Contested

It is highly likely that above-mentioned instabilities increase uncertainty about the future among economic actors and policy makers. Uncertainty, Knight (1921) argued, refers to situations in which economic actors cannot make probability calculus whereas they can in case of risk. His aim was not to define risk and uncertainty but to explain the origin of profit in capitalist economies, which is similar to Schumpeter's (1934) explanation.

Neoclassical economists have always strongly opposed to the distinction between risk and uncertainty, which they consider as meaningless. Not surprising because uncertainty radically conflicts with rational choice theory, the cornerstone of neoclassical economics, and with their deterministic view that the future is in principle knowable. Expectations of the future are, in their view, the product of rationality, i.e. they are based on rational calculations of probable outcomes associated with various possible choices. Lucas' (1972) theory of rational expectations is the most sophisticated formulation of the idea of rationality of expectations. Neoclassical economists argue that probabilities on future events can *always* be calculated. Knight's distinction between risk and uncertainty, they argue, is based on objective probability calculus (the numerical probability of relative frequencies of events in the past) only. But uncertainty can be quantified on the basis of subjective assessments of probability (the probability of the degree of belief or feeling of (un)certainty). By integrating the theory of subjective probability, rational choice, and expected utility (an idea that incorporates the utility of each outcome and its probability of its occurrence), the apparent dichotomy between subjective and rational belief has, in their view, been overcome, and the distinction between risk and uncertainty is meaningless (Köhn 2017, 63–94).

The heterodox economist Shackle (1979) argued against neoclassical economists that probability calculus is meaningless in economics. Economic actors can make real choices that are not determined by the past but can neither foresee the consequences of their own choices nor those of others. In addition, dynamic capitalism is full of novelties which is why the conditions of probability calculus cannot be fulfilled. Frequency probabilities require a large number of trials under equal conditions in time and space. In many situations, there are enough pooled results from which statistical probability of risk can be derived, such as in the case of, for instance, house fires. But economic actions alter the conditions under which they take place which makes it impossible to determine probabilities. Once entrepreneurs have made investments in a specific place, the whole (spatial) field of economic relations has changed, so that it becomes impossible to take the same decision again on the same conditions. Due to these economic dynamics, it is impossible to know ex ante all possible consequences of a decision. The impossibility of knowing an exhaustive list of possible consequences precludes any objective probability calculus because it is based on the assumption that all possible consequences are known a priori. The same applies to subjective probability calculus in which degrees of belief are expressed in numerical terms. Situations of uncertainty, Shackle argues, are characterized by a lot of possibilities of surprise, so that neither the consequences of an action can be anticipated nor can a comprehensive list of expectations and its consequences be made. Possible consequences can only be imagined.

From his arguments, one should not conclude that future in dynamic capitalism is a totally unknowable "void", Shackle (1979) seems to believe. Uncertainty has to be reduced in one way or another because neither markets nor politics can function under radical uncertainty. Some predictability is a precondition for their functioning. In a world of uncertainty, institutions, habits, firm routines, norms, and trust play a functional role in providing a basis for decision-making (Hodgson 1988), and so do fictional expectations, as said before. Capitalism is characterized by a mixture of order and change (Beckert 2009), and the future is almost always a messy mixture of persistent regularities of behavior and stable institutional or physical constraints on the one hand and novelty and innovation on the other. In addition, "we can often discern pointers in the present to patterns that may later unfold, and make pattern predictions about the future in complex and creative systems, based on what we do know about relatively stable constraints and incentives" (Beckert and Bronk 2018, 10).

5.6 Fictional Expectations as a Motivational Force of Action

Entrepreneurs who plan to make an investment or to start a new firm cannot make a probability calculus of the future revenues of their investments because they do not know the future behavior of others and, more generally, their business environment is dynamic. Because any probability calculus of their future state of affairs is impossible, the only thing they can do to cope with uncertainty is to act on the basis of imaginaries and expectations regarding their future revenues in relation to all factors that are expected to influence them. These imaginaries and expectations give orientation to their process of decision-making, despite the incalculability of the outcomes. Beckert (2016) calls these expectations "fictional". They refer to "the images actors form as they consider future states of the world, the way they visualize causal relations, and the ways they perceive their actions in influencing outcomes" (Beckert 2016, 9). Thus, it refers to images of the future that economic actors create and share with one another, and that shape and motivate present-day economic action. Beckert calls the adjective of expectations fictional to express that images of the future are a kind of fiction. Drawing on the work of philosophers and literary critics, he gives convincing arguments that economic imaginaries and narratives share formal features with fictional narratives in literary fiction, although there are differences (Beckert 2016, 67–75). By creating economic images and expectations economic actors create a reality of its own and act as if it were true, just as authors and readers of literary fiction act as if the described reality were true. Fictional expectations about the future do not need to be true (they cannot be true) but have to be convincing to be effective. Only then actors will act in accordance with fictional expectations. Convincing fictional expectations provides justifications for decisions whose success is uncertain. Business confidence in future economic growth is a case in point. Akerlof and Shiller (2009) distinguish, in addition to the well-known consumption-, investment-, and government expenditure multiplier, a confidence multiplier that represents the change in income that results from a one-unit change in confidence, however, it might be conceived or measured. Changes in confidence investment opportunities or future economic growth will result in changes in income and confidence in the next round, and each of these changes will in turn affect income and confidence in yet further rounds.

While imagined futures and fictional expectations are expressed by individuals, they are, to a large extent, social. The imaginaries and fictional expectations of consumers who plan to book a holiday trip are shaped by their class- and lifestyle position, personal networks, media (pictures, movies, opinion leaders, and influencers), and advertisements. And the imaginaries and expectations of entrepreneurs are also social because they are shaped by their past and present experiences (see below), the sector they belong to (and having a dominant position within a sector or not), their inter-firm networks, business associations, banks, consultants, and the state. Shared imaginaries and fictional expectations form a collective belief which makes them more effective in coordinating economic actions. It simultaneously makes them an object of power struggle. Powerful actors (businessmen, business associations, politicians, experts, and media) try to influence the construction of imagined futures and the alleged causal relations that underpin them according to their interests. They try to frame discourses about imagined futures and dominate narratives about the future.

Forecasting techniques (forecasts of market development, anticipation of competitors' strategies, macro-economic forecasting, prognoses of prices of financial assets, and scenarios) help form fictional expectations about the future systematically. Actors (firms, research institutes, and government agencies) pretend that their forecasts are grounded on probability calculations and are often represented as objective and rational extrapolations which is, as argued before, impossible under conditions of uncertainty; and predictions of the future often turn out to be completely wrong. Yet, these forecasting techniques are performative, in the sense that actors behave and coordinate their actions accordingly (see for more Beckert and Bronk 2018).

All above-mentioned factors that shape imaginaries and fictional expectations raise the question of what the role of history and geography is in shaping them. The fact that actors cannot form rational expectations on the basis of past data under conditions of uncertainty does not mean that fictional expectations of the future are not influenced nor even largely shaped by the past.

As argued before, actors simultaneously oriented toward the past, the future, and the present at any given moment, although they may be primarily oriented toward one of them in specific situations. Thus, the same social factors that shape imaginaries and fictional expectations about the future are themselves, to a large extent, the outcome of the present and the past. "Expectations in the economy are not free-floating fantasies: outlooks on the future and the courses of action that are based on them are socially constrained through the distribution of wealth and power, through cognitive frames, through networks, through formal and informal institutions, and through normative obligation" (Beckert 2016, 60). Thus, imaginaries and fictional expectations are, to some extent, path-dependent but how much is an open question (see for this point Guseva and Mooney 2018).

A similar argument can be made with respect to geography, which is completely ignored in discussions on imagined futures and fictional expectations. If assessments of the future are influenced by past experiences and their interpretation, as well as established structures, then they vary not only in time but also in space. Regions follow a path-dependent development path. If fictional expectations are influenced by established structures, past experiences and their interpretation, then it is highly plausible to assume that they vary from place to place according to the development path they have followed, i.e. their industrial structure (specialized or diversified, "old" and "new sectors"), being booming or in crisis, and the scaled combination of national and local forms of capitalism regarding inter-firm- and state-business relations, and capital-labor relations among others. In short, there are good reasons to believe that the imaginaries and fictional expectations of the future of entrepreneurs are place-based.

5.7 Bringing Fictional Expectations in Regional Economic Resilience

As mentioned before, a high adaptability and innovativeness in "normal" times makes regional economies less vulnerable to shocks and more successful in recovering from them. In "normal" times actors anticipate to the future and form expectations about them and may proactively change their behavior accordingly. Probably most of the time firms, whether or not in collaboration with business associations and policy makers, innovate and adapt to changing circumstances without anticipating to possible shocks. They merely expect to strengthen their competitive position without the intention to be more shock proof during a shock. However, they sometimes do anticipate to possible shocks and act accordingly in "normal" times with the intention to be more shock proof in the future. A good example of anticipatory adaptation is the Port Authority of Rotterdam (Havenbedrijf Rotterdam) (Section 5.7.1).

Actors do not only anticipate to possible shocks but also respond to them. They can change their activities or even location, and can transform their behavior strategically through innovations or otherwise. How precisely they respond to a shock depends, to a large extent, on their future expectations that are shaped by the past and the present. In this section, we explore on the basis of interviews the future expectations of tourism firms, tourism business associations, and policy makers in Antalya during the deep COVID-19 crisis in 2020 and 2021 that severely hit the city (see for details Erkuş and Terhorst 2023) (Section 5.7.2).

5.7.1 Anticipatory Adaptation of the Port Authority of Rotterdam to Possible Shocks

The port of Rotterdam is Europe's largest port industrial complex with oil refineries, (petro)chemical industries, oil- and natural gas storage facilities, electricity plants, container terminals, and terminals for dry-bulk (coal, iron ore, and wheat) and wet-bulk (oil, gasoline, and, very importantly since Russia's strong reduction of gas supplies, liquefied natural gas, imported from Norway and Qatar). The Port Authority of Rotterdam, a limited liability company the shares of which are owned by the city of Rotterdam (70%) and national state (30%), is mainly responsible for port infrastructure investments, restructuring of older port areas, and the zoning of new land reclaimed from the North Sea on behalf of port development (the Maasvlakte).

The port of Rotterdam faces uncertainty about which investments have to be made in the future to attract new value chains, new economic activities, and clean energy carriers. These investment decisions are dependent on the following geopolitical, economic, technological, and societal developments in the future. Will the world make a radical energy transition in the coming decades or will it go slowly or even stagnate? Will global free trade continue to flourish or will the dominant power blocs US and the EU aiming to lessen their economic and political dependency on China erect trade barriers to protect their own industries? Will consumers prefer high quality to price or the other way around? To get some grip on those uncertainties the Port Authority made four scenarios (Port of Rotterdam Authority 2022).

1 Connected deep green: good global cooperation with an acceleration of digital transparency in logistics and very active worldwide measures against climate changes resulting in global CO_2-neutrality in 2050, a maximum rise of temperature of 1.5°C, a high rate of economic growth and a fairly equal level of living. The impact of this scenario on the port of Rotterdam is a substitution of all fossil industries (the petro-chemical complex!) for renewable energy such as hydrogen (the Port Authority has already made the first steps in becoming a hub hydrogen storage).
2 Regional well-being: around 2030 regional clusters of countries start collectively with ever more focusing on a clean and healthy living environment, privacy, and well-being due to a lack of global trust. This leads to a deteriorating business climate for basic manufacturing industries and a moderate

rate of economic growth. This means for the port of Rotterdam a strong decline of crude oil, coal, and iron ore because of the decline of fossil energy industries but a growth of the volume of breakbulk because of the growth of semi-finished products.

3 Protective markets: the major global power blocs mistrust each other accompanies with geopolitical tensions and suboptimal integration of logistic value chains. In such a fragmented world economic interests conflict with each other leading to a focus on national or supra-national self-reliance, financial prosperity, resilience, and defense. No CO_2 neutrality before 2100 and low economic growth. Trade restrictions lead to a slow transition to renewable energy. In the port mixed cargo will decline due to re- and nearsshoring and oil refining will strongly decline.
4 Wake-up call: increasing worries about economic consequences of external shocks like the availability of energy and food or extreme weather generate a turning point. Strategic cooperation and rigorous measures are perceived to be required to reduce CO_2-emissions. This leads to strategically strong EU policy measures, moderate economic growth, and a late but quick transition to sustainable energy. This means for the port a growth of container transport and biomass as raw material for energy plants and chemical industries. The late but quick energy transition requires the collection of CO_2 emissions from industries in order to be transported by pipeline to empty gas fields in the North Sea (this project starts in the near future).

The Port Authority of Rotterdam also faces another uncertainty about the future, namely a rising sea level due to climate change. Much of western and northern parts of the Netherlands are located below sea level but are protected from floods by dikes. Although the land of the port of Rotterdam is generally higher than land in adjacent areas and is protected from floods by the Maeslant barrier that can be closed at very high tide, it is not surrounded by dikes. Thus, there is a real chance that they will be flooded with a rising sea level in combination with heavy storms. It goes without saying that floods will cause a huge economic and environmental damage (damaged oil refineries, chemical plants, and broken oil- and gas pipelines). Not all port subareas are equally vulnerable to floods, however. Some are higher located than others. To anticipate to such a shock the Port Authority has formulated various adaptation strategies with cost-benefit analyses in the future that vary by port subarea and type of economic activity: in some cases, raising land with fluid sand or building dikes and in other cases making assets and installations more waterproof (RoyalHasoningDHV 2022).

5.7.2 Fictional Expectations of Post-COVID-19 Tourism in Antalya

Antalya is a sun-sea-sand mass-tourism city with a tourism economy that is mainly dependent on foreign tourists (particularly from Russia, the UK, Germany, and the Netherlands). Over the last twenty years, it has followed a very dynamic development path with a lot of adaptations and innovations. The city's tourism economy has been fairly resilient to crises like the birds flu in 2003, the global financial crises of 2008, and Turkish (geo)political crisis in the period 2014–2018 (Terhorst and Erkus-Öztürk 2019). This is in line with the region's adaptability and innovativeness. Its tourism economy has been strongly diversified over the last twenty years (Erkuş-Öztürk and Terhorst 2018), and many restaurants and hotels have been fairly innovative (Erkuş-Öztürk and Terhorst 2016).

During the COVID-19 period, the business confidence of all tourism firms, representatives of business associations, and policy makers was very high. They all believe that Antalya's tourism economy will prosper again in the post-COVID-19 era, particularly on the basis of high quality for low prices. This does not mean that all their other expectations of the future are the same. They are not. Some interviewees have a static view on Antalya's tourism economy and their expectations are implicitly based on an engineering conception of regional economic resilience. They expect business as usual in the future, i.e. the European and Russian markets will remain as dominant as before, and they are blind to the environmental impact of flight tourism and do not expect that the environmental consciousness of foreign tourists will have any effect on the development of tourist arrivals. Most other interviewees, by contrast, have a dynamic view on Antalya's tourism economy. They have implicitly a somewhat adaptive conception of resilience in their mind. But their expectations differ regarding the role the COVID-19 crisis plays. Some interviewees point to the adaptability of Antalya's tourism economy in the past: it quickly bounced back from earlier crises and flexibly responded to new developments in the past. They expect and advocate more diversification of the city's tourism economy and believe that it can easily respond to environmental policy changes in the future (although they do not expect strict environmental policy measures). But their expectations of the future are not triggered by the COVID-19 crisis: they expect tendencies that have been started (long) before the crisis. Other interviewees, however, expect a lot of changes that are triggered by the COVID-19 crisis and have already made decisions on the basis of their expectations. They expect (and some have already acted accordingly) a strong growth of Turkish higher-class tourists, more environmentally friendly hotels with a lot of social distance possibilities in isolated places, more luxurious villa- and caravan tourism with private pools, more B-leisure, staycations, clamping, and hybrid hotels. In addition, they expect a growing importance of self-reservation systems that will undermine the market dominance of tour operators.

5.8 Conclusion

It is no exaggeration to say that social scientists are generally more focused on the present and the past than on the future. Actors who have to make decisions in radical uncertainty about the future cannot do otherwise than to form fictional expectations and imaginaries of the future to motivate and legitimate their actions, a point that has, until recently, been ignored by researchers of regional economic resilience.

Although the call for more agency in analyzing regional economic resilience has become louder, it is seldom specified what it precisely means. As we have seen, actors who make judgments in the present are always simultaneously oriented to the past and the future, be it in varying degrees. That means that fictional expectations about the future actors form in the present are rooted in and shaped by the structure of the past. The windows of opportunity through which actors look at the future are not infinitely wide and open but are limited by the structural constraints of the past, as illustrated in our two case studies that show an adaptation to (possible) shocks.

Policy makers of the Port Authority of Rotterdam anticipate in a non-shock period to possible future shocks generated by geopolitical conflicts and the rise of the sea water level. They try to form fictional expectations by means of scenarios on behalf of their infrastructure investment planning in the future. They aim to maintain the core function of the port by reorienting and reorganizing its structure. Thus, all scenarios are based on the industrial composition of the port, its strategic position in the Dutch economy, and its huge public and private investments of the past. Maybe these scenarios reduce the uncertainty of the policy makers somewhat but not very much. In some scenarios, the power of some big multinationals like ECT, Hutchison, Maersk, and Shell will be reduced but their fictional expectations are not included in the scenarios. Thus, there is no grand narrative supported by all key actors in the port of Rotterdam to cope with geopolitical tensions and the rise of the sea water level.

The tourism city of Antalya was severely hit by the COVID-19 pandemic. But the future expectations of all key actors in its tourism industry were nevertheless very positive which were largely based on the past: the high growth of the city's tourism industry before the pandemic, its quick recovery from former crises, and its adaptability to new developments by ever more diversification. This is the narrative told by all. Some actors expected that the COVID-19 crisis would trigger a lot of changes in the city's tourism industry and have already acted accordingly. They come down to more diversification. Thus, the COVID-19 crisis has already reinforced the city's tourism development path toward more diversification.

References

Akerlof, G. A., and R. J. Shiller. 2009. *Animal Spirits. How Human Psychology Drives the Economy, and Why it Matters for Global Capitalism*. Princeton: Princeton University Press.

Bashkar, R. 1979. *On the Possibility of Naturalism. A Philosophical Critique of the Contemporary Human Sciences*. Brighton: The Harvester Press.

Beckert, J. 2009. The Social Order of Markets. *Theory and Society* 38 (3):245–69. DOI: 10.1017/9781108557626

Beckert, J. 2016. *Imagined Futures, Fictional Expectations and Capitalist Dynamics*. Cambridge, MA: Harvard University Press.

Beckert, J., and R. Bronk, eds. 2018. *Uncertain Futures. Imaginaries, Narratives, and Calculation in the Economy*. Oxford: Oxford University Press.

Boschma, R. 2015. Towards an Evolutionary Perspective on Regional Resilience. *Regional Studies* 29 (5):733–751. DOI: 10.1080/00343404.2014.889579

Brakman, S., H. Garretsen, and C. Van Marrewijk. 2015. Regional Resilience across Europe: on Urbanisation and the Initial Impact of the Great Recession. *Cambridge Journal of Regions, Economy and Society* 8 (2):225–40. DOI: 10.1093/cjres/rsu022

Bristow, G., and A. Healy. 2014a. Regional Resilience: An Agency Perspective. *Regional Studies* 45 (5):923–35. DOI: 10.1080/00343404.2010.486503

Bristow, G., and A. Healy. 2014b. Building Resilient Regions: Complex Adaptive Systems and the Role of Policy Intervention. *Raumforschung und Raumordnung* 72 (2):93–102.

Bristow, G., and A. Healy. 2020. Regional Economic Resilience: an Agency Perspective. In *Handbook on Regional Economic Resilience*, ed. G. Bristow and A. Healy, 36–53. Cheltenham: Edward Elgar. DOI: 10.4337/9781786438979

Capello R., A. Caragliuy, and U. Fratesi. 2015. Spatial Heterogeneity in the Costs of the Economic Crisis in Europe: Are Cities Sources of Regional Resilience? *Journal of Economic Geography* 15 (5):951–972. DOI: 10.1093/jeg/lbu025

Cowell, M. M. 2013. Bounce Back or Move on: Regional Resilience and Economic Development Planning. *Cities* 30:212–222.

Crescenzi, R., D. Luca, and S. Milio. 2016. The Geography of the Economic Crisis in Europe: National Macroeconomic Conditions, Regional Structural Factors and Short-term Economic Performance. *Cambridge Journal of Regions, Economy and Society* 9 (1):13–32. DOI: 10.1093/cjres/rsv028

Davies, S. 2011. Regional Resilience in the Downturn: Comparative Evidence from European Countries. *Cambridge Journal of Regions, Economy and Society* 4 (3):369–82. DOI: 10.1093/cjres/rsr021

Emirbayer, M., and A. Mische. 1998. What is agency? *American Journal of Sociology* 103 (4):962–1023. DOI: 10.1086/231294

Eraydin, A. 2016. The Role of Regional Policies along with the External and Endogenous Factors in the Resilience of Regions. *Cambridge Journal of the Regions, Economy and Society* 9 (1):217–34. DOI: 10.1093/cjres/rsv028

Erkuş-Öztürk, H., and P. Terhorst. 2016. Innovative Restaurants in a Mass-Tourism City: Evidence from Antalya, *Tourism Management* 54:477–89. DOI: 10.1016/j.tourman.2015.12.007

Erkuş-Öztürk, H., and P. Terhorst. 2018. Economic Diversification of Single-Asset Tourism City: Evidence from Antalya. *Current Issues in Tourism* 21 (4):422–39. DOI: 10.1080/13683500.2016.1261087

Erkuş, H., and P. Terhorst. 2023. Imagined Futures of Post-Covid 19 Tourism in Antalya. *Advances in Hospitality and Tourism Research (AHTR). An International Journal of Akdeniz University Tourism Faculty* 11 (1):171–90. DOI: 10.30519/ahtr.1112787

Evans, R., and J. Karecha. 2014. Staying on Top: Why is Munich so Resilient and Successful? *European Planning Studies* 22 (6):1259–79. DOI: 10.1080/09654313.2013.788883

Evenhuis, E. 2017. New Directions in Researching Regional Economic Resilience and Adaptation. *Geography Compass* 11 (11):1–15. DOI: 10.1111/gec3.12345

Fuller, C. 2023. Uncertainty, fictional expectations and economic agency. *Geoforum* 140:103699. DOI: 10.1016/j.geoforum.2022.103699

Giannakis, E., and A. Bruggeman. 2017. Determinants of Regional Resilience to Economic Crisis: a European Perspective. *European Planning Studies* 28 (8):1394–415. DOI: 10.1080/09654313.2017.1342345

Grillitsch, M., and B. T. Asheim. 2023. Towards Regenerative Regional Development in Responsible Value Chains: An Agentic Response to Recent Crises. *European Planning Studies*. DOI: 10.1080/09654313.2023.2205890

Grillitsch, M., and M. Sotarauta. 2020. Trinity of Change Agency, Regional Development Paths and Opportunity Spaces. *Progress in Human Geography* 44 (4):704–23. DOI: 10.1177/0309132519853870

Guseva, A., and H. Mooney. 2018. Time, role of the past and varieties of fictional expectations: Comments on Jens Beckert's Imagined Futures. *Distinktion. Journal of Social Theory* 19 (3):336–40. DOI: 10.1080/1600910x.2018.1502089

Hodgson, G.M. 1988. *Economics and Institutions. A Manifesto for a Modern Institutional Economics*. Cambridge: Polity Press.

Hudson, R. 2010. Resilient regions in an uncertain world: Wishful thinking or a practical reality? *Cambridge Journal of Regions, Economy and Society* 3:11–25. DOI: 10.1093/cjres/rsp026

Jolly, S., M. Grillitsch, and T. Hansen. 2020. Agency and Actors in Regional Industrial Path Development. A Framework and Longitudinal Analysis. *Geoforum* 111:176–88. DOI: 10.1016/j.geoforum.2019.12.013

Kakderi, C., and A. Tasopoulou. 2017. Regional Economic Resilience: the Role of National and Regional Policies. *European Planning Studies* 25 (8):1435–53. DOI: 10.1080/09654313.2017.1322041

Knight, F. H. 1921. *Risk, Uncertainty and Profit*. Charleston: Nabu Press.

Köhn, J. 2017. *Uncertainty in Economics, A New Approach*. Cham: Springer. DOI: 10.1007/978-3-319-56077-9

Lucas, R. E. Jr. 1972. Expectations and the Neutrality of Money. *Journal of Economic Theory* 4 (April):103–24. DOI: 10.1016/0022-0531(72)90142-1

Magro, E., E. Uyarra, and J. M. Valdaliso. 2022. Agency, Institutions and Regional Resilience: An Approach from the Basque Region. In *Towards Resilient Organizations and Societies. A Cross-Sectoral and Multi-Disciplinary Perspective*, ed. R. Pinheiro, M. L. Frigotto, and M. Young, 277–303. Cham: Palgrave MacMillan/Springer Nature.

Martin, R. 2012. Regional Economic Resilience, Hysteresis and Recessionary Shocks. *Journal of Economic Geography* 12 (1):1–32. DOI: 10.1093/jeg/lbs033

Martin, R. L. 2018. Shocking Aspects of Regional Development: Towards an Economic Geography of Resilience. In *The New Oxford Handbook of Economic Geography*, ed. G. L. Clark, M. P. Feldman, M. S. Gertler, and D. Wójcik, 839–871. Oxford: Oxford University Press.

Martin, R., and P. Sunley. 2015. Towards a Developmental Turn in Evolutionary Economic Geography. *Regional Studies* 49 (5):712–32. DOI: 10.1080/00343404.2014.889579

Martin, R., and P. Sunley. 2020. Regional Economic Resilience: Evolution and Evaluation. In *Handbook on Regional Economic Resilience*, ed. G. Bristow and A. Healy, 10–35. Cheltenham: Edward Elgar Publishing.

Marx, K. [1852] 1995. The Eighteenth Brumaire of Louis Bonaparte. www.marxists.org.

Pike, A., S. Dawley, and J. Tomaney. 2010. Resilience, Adaptation and Adaptibility. *Cambridge Journal of Regions, Economy and Society* 3 (1):59–70. DOI: 10.1093/cjres/rsq001

Port of Rotterdam Authority. 2022. Scenarios for 2050, White Paper. Accessed April 1, 2024. https://www.portofrotterdam.com/sites/default/files/2022-12/white-paper-future-scenarios-2050_0.pdf

RoyalHasoningDHV. 2022. *Een Waterveilige Rotterdamse Haven. Een Adaptatiestrategie voor een Veilige Haven - Nu en in de Toekomst*. [A Water-Safe Port of Rotterdam. An Adaptation Strategy for a Safe Port - Now and in the Future]. Nijmegen: HaskoningDHV Nederland.

Sotarauta, M., and M. Grillitsch. 2023. Studying Human Agency in Regional Development, *Regional Studies* 57 (8):1409–14. DOI: 10.1080/00343404.2023.2217224

Scott, A. J. 2012. *A World in Emergence: Cities and Regions in the 21st Century*. Cheltenham: Edward Elgar.

Schumpeter, J. A. 1934. *The Theory of Economic Development. An Inquiry into Profits, Capital, Credit, Interest and the Business Cycle*. London: Oxford University Press.

Schumpeter, J. A. [1942] 1976. *Capitalism, Socialism and Democracy*. London: Allen & Unwin.

Shackle, G. L. S. 1979. *Imagination and the Nature of Choice*. Edinburgh: Edinburgh University Press.

Sutton, J., A. Arcidiacono, G. Torrisi, and R. N. Arku. 2023. Regional Economic Resilience; a Scoping Review. *Progress in Human Geography* 47 (4):500–532. DOI: 10.1177/03091325211 06957

Sutton, J., and G. Arku. 2022. Regional Economic Resilience: towards a System Approach. *Regional Studies, Regional Science* 9 (1):497–512. DOI: 10.1080/21681376. 2022.2092418

Tan, J., X. Hu, R. Hassink, and J. Ni. 2020. Industrial structure or Agency: What Effect Regional Economic Resilience? Evidence from resource-Based Cities in China. *Cities* 106:102906.

Terhorst, P., and H. Erkus-Öztürk. 2019. Resilience to the Global and Turkish (Geo) Political Crisis Compared. *Tijdschrift voor Economische en Sociale Geografie* 110 (2):138–55. DOI: 10.1111/tesg.12343

Tooze, A. "chartbook#130, Defining Polycrisis-from Crisis Pictures to the Crisis Matrix.", June 24, 2022, https://adamtooze.substack.com/

Völlers, P., T. Neise, P. Verfürth, M. Franz, F. Bücken, and K. P. Schumacher. 2023. Revisiting Risk in the Global Production Network Approach 2.0 - Towards a Performative Risk Narrative Perspective. *Environment and Planning A: Economy and Space*. Advance online publication. DOI: 10.1177/0308518X231169288

Section B

Economic Transition and Resilience of Companies and Regions

6 Learning from the COVID-19 Crisis – How Can Business Development and City Marketing Agencies Support the Restaurant and Bar Industry?

Martin Franz, Thomas Neise, Philip Verfürth, and Philip Völlers

6.1 Introduction

The COVID-19 pandemic had a strong impact on the restaurant and bar industry (e.g. Brizek et al. 2021; Hakim, Zanetta, and da Cunha 2021; Neise, Verfürth, and Franz 2021): businesses had to close temporarily and even if they were allowed to be open, guests preferred to stay away, hygiene requirements increased costs, staff dropped out due to illness, amongst other challenges. In Germany, this crisis occured in a situation in which processes of change on the structural and format level have already led to the closure of many establishments before the pandemic (Franz 2020b). Yet, it was only the shock of the COVID-19 pandemic that led many municipalities to recognize the importance of this industry for the attractiveness of cities, their local economy, and the need for action to support the industry in Germany. In this context, this chapter addresses four questions, which also structure the chapter:

1. Why should business development institutions and city marketing agencies be concerned with the restaurant and bar industry?
2. What were the pre-existing changes and challenges experienced by this industry before the onset of the COVID-19 pandemic?
3. How does the COVID-19 pandemic affect this economic sector?
4. What strategies can business development and city marketing agencies employ to enhance the resilience of the local restaurant and bar industry during crisis situations?

To answer these questions, a case study of Germany's restaurant and bar industry is employed. The empirical basis of the chapter consists of a quantitative survey (n = 623) and 49 qualitative interviews with managers, owners, and experts in the field of restaurants and bars – including representatives of the German Hotel and Restaurant Association (DEHOGA). Thirty-three of these interviews were conducted before the COVID-19 pandemic in 2018–2019.

DOI: 10.4324/9781003457626-8

Additional 16 interviews were conducted in 2020–2023. The interviews were evaluated according to qualitative content analysis (Mayring 2014). Quotes from the interviews were translated from German into English for this chapter.

The quantitative survey was conducted in the form of an online questionnaire. The questionnaire includes information on responses to the COVID-19 pandemic, use of government assistance, pre-crisis business performance, and general business characteristics. The survey was distributed through business associations, such as the regional branches of the Chambers of Industry and Commerce (IHK), the DEHOGA, and local and regional business development institutions in Germany. A total of 623 respondents participated between April 19 and June 10, 2020. In total, 36.6 percent of businesses in the survey sample are located in rural areas. Respondents were key decision-makers primarily owners (81.5 percent) and managers (15.5 percent). In total, 32.4 percent of the respondents were female, and 67.2 percent male. The survey respondents were, on average, 50 years old.

6.2 Organizational Resilience in the Restaurant and Bar Industry

Economic-geographical studies on resilience mainly focus on the concept of regional resilience adopting a meso-level perspective. Individual companies (micro-level) as key agents of building resilience capacities in a region are somewhat understudied (e.g. Billington et al. 2017; Neise, Verfürth, and Franz 2021; Verfürth et al. 2022). Accordingly, this chapter is based on an understanding of resilience oriented on the approach of so-called organizational resilience. Organizational resilience can be defined as "a firm's ability to effectively absorb, develop situation-specific responses to, and ultimately engage in transformative activities to capitalize on disruptive surprises that potentially threaten the firm's survival" (Lengnick-Hall, Beck, and Lengnick-Hall 2011, 244). In this context, the resilience of companies in a crisis is understood as the result of current adaptations and cumulative previous experience from historical structural changes and transformations (Tsiapa and Batsiolas 2018, 2).

To answer the question of what business development and city marketing agencies can do to strengthen the resilience of restaurants and bars, it has to be clarified which factors influence the resilience of companies in the first place. It is necessary to differentiate between a company's short-term ability to cope with crises and its ability to develop appropriate long-term adaptation strategies retrieved from prior crisis experiences. While the literature on crises in the restaurant and bar industry was minimal before the COVID-19 crisis, there is more extensive literature on the success factors and reasons for closing restaurants and bars (e.g. Le 2015; Parsa et al. 2011; Schneider 2008). During the COVID-19 crisis, several research studies were conducted on the impact of restaurants and bars, but due to the topicality of the issue, they only looked at short-term effects and mitigation strategies (e.g. Matsenko et al. 2021; Mitrofanova, Chernova and, Batmanova 2022; Neise, Verfürth, and Franz 2021; Verfürth et al. 2022). Therefore, it is important to combine the literature on the success factors and reasons for closing restaurants and bars and the literature on the impacts of COVID-19 on restaurants and bars with the literature on business resilience.

Coping measures are reactive and incremental measures that aim to minimize the negative impact of shocks (Neise and Revilla Diez 2019). Tibay et al. (2018) found that restaurants and bars in Auckland are responsive to slow changes. However, there is a knowledge gap concerning coping measures in place in the event of major shocks. This is because natural disasters or other shocks rarely occur. Hence, it is necessary to emphasize the implementation of coping measures (Tibay et al. 2018) because they can strengthen the competitiveness of companies not just in the context of shocks and crises.

Coping measures companies take are diverse, ranging from cost reductions and improved marketing to (temporary) closure (Biggs, Hall, and Stoeckl 2012). The impact of the measures on companies' resilience depends largely on the point in time of the observation. This is illustrated by the example of staff layoffs in crises. While layoffs may contribute to survival in the short term, they may affect companies' resilience in the long term (Gittell et al. 2006).

For a company's long-term success, it is crucial to develop suitable adaptation strategies from prior crisis experiences (Berman, Quinn, and Paavola 2012). In contrast to coping, adapting to crises is often accompanied by structural changes, such as realigning the business model and diversifying the product range (Li et al. 2021). During the COVID-19 crisis, for example, many businesses introduced delivery services (e.g. Crîşmariu et al. 2022). Companies with delivery and pickup services experienced a smaller drop in sales than those that did not use these services (Franz, Müller, and Neise 2020). Studies illustrate that particularly proactive adaptation strategies have a positive impact on the resilience of companies. This makes organizational decision-makers "active agents in resilience, and resilience [...] a set of [response] capacities that can be intentionally developed and acquired" (Bristow and Healy 2014, 929) by individual owners and managers of restaurants and bars (cf. Boschma 2015). As shown by Alonso-Almeida and Bremser (2013), using the case of the Spanish hospitality sector in Madrid, companies proactively entering new market segments and investing in IT infrastructure were more competitive one year after the 2008 global financial crisis than companies with reactive adaptation strategies.

Research highlights that subjective experiences, perceptions, and knowledge of individual decision-makers and company characteristics as well as capabilities affect organizational resilience. Various studies show that the education and professional experience of company owners and managers, as well as company size and age, have a positive influence on organizational resilience (Headd 2003; Kaniovski and Peneder 2008; Khan and Sayem 2013; Parsa et al. 2005; Parsa et al. 2015; Neise, Verfürth, and Franz 2021). In addition, critical company-level characteristics and capabilities are identified: an important indicator of whether companies can withstand a crisis is their pre-crisis profitability. If companies have already had problems in this respect, they are more likely to be unable to resist a shock (Esteve-Pérez and Mañez-Castillejo 2008). In Germany's restaurant and bar industry, many companies were already in a difficult financial situation pre-pandemic, triggered by changes on the demand side and in the business environment due to a competitive market situation and regulatory pressure (Franz 2020b; Neise, Verfürth, and Franz 2021). However, for companies to demonstrate strong resilience, it is a prerequisite that they have resources (e.g. technological, human, and financial) to sustain business performance during a crisis (Conz and Magnani 2019). Parsa et al. (2011, 374) note that "[...] as restaurants get larger and more complex, more resources are likely to be used (financial and human), and there is a greater chance of survival. Relatively small and simple operations, requiring fewer resources and managerial expertise, appear more vulnerable to failure". In addition to financial capital, real estate ownership is essential – especially for the restaurant industry (Neise, Verfürth, and Franz 2021).

Although research considers individual and company-specific characteristics and capabilities, recent studies increasingly emphasize the importance of external factors for resilience in crisis situations. For example, some studies examine the influence of location factors (Andres and Round 2015; Scuderi, Tesoriere, and Fasone 2020) and regional economic factors (Tsiapa and Batsiolas 2018) on companies' resilience. This also includes access to resources provided by extra-organizational formal and informal relationships with actors and shareholders (Stevenson 2014). This is because managing crises and adapting strategies often involve coordinating with other businesses, industries, civil society, and political bodies (Linnenluecke and Griffiths 2015; Neise, Revilla Diez, and Garschagen 2018). In addition to material resources, access to intangible resources such as social capital, knowledge, and information can help companies cope with crises and develop adaptation strategies. For example, Marcos and Macaulay (2008) argue that companies' external relationships act as sensors to provide information about emerging environmental changes. This information can positively influence companies' adaptive capacity (Berkhout 2012). Accordingly, previous studies (e.g. Martin 2018) suggest that supportive action of external actors and organizations, such as business development and city marketing agencies, are a critical factor in determining whether companies are resilient in crisis situations or not.

6.3 Why Should Business Development and City Marketing Institutions Deal with the Restaurant and Bar Industry?

For decades, the restaurant and bar industry has been largely neglected by business development, city marketing, and urban development agencies in Germany. Changes in the restaurant and bar industry were often perceived as a symbol or consequence of developments in urban development, but not as a field of work that planning, politics, and business development agencies should actively address (Pätzold 2012, 2014; Pohl 2010). For example, in 2018, the executive director of the city marketing institution of a major northern German city said in an interview about promoting the restaurant and bar industry, "I don't see any need there. It runs by itself". As in many city marketing institutions, her focus was on tourism and retail.

In an online survey of business development agencies representatives conducted before the COVID-19 pandemic, 39 percent of respondents said they were already addressing night-time economy issues, and another 19 percent planned to do so in the near future. Forty-two percent of the business development agency staff stated that they were not planning to become involved in the night economy, nor would they do so in the future (see Figure 6.1; Konermann 2022, 148). However, the night-time economy includes the restaurant and bar industry but also, for example, cultural offerings (on the night-time economy, see e.g. Krüger, Schmid, and Jauernig 2015; Schmid 2018).

The relatively small interest shown by business development and city marketing agencies in the night-time economy, and in particular in the restaurant and bar industry, is in stark contrast to the major role that these sectors play in a city's image, appeal, and attractiveness. Especially in times when stationary (brick-and-mortar) retail is losing importance for the customers' decision to visit cities or neighborhoods, the relative significance of the restaurant and bar industry in this respect is increasing. Restaurants, and outdoor catering in particular, potentially increase the sojourn quality – much invoked in connection with the attractiveness of city centers (e.g. Baier and Engelen 2015; Pätzold 2014; Stepper 2015).

Florida's books (e.g. 2005) on creative cities and the so-called creative class have raised awareness in academia and practice of the importance of an attractive restaurant and bar offer as a soft location factor for attracting professionals

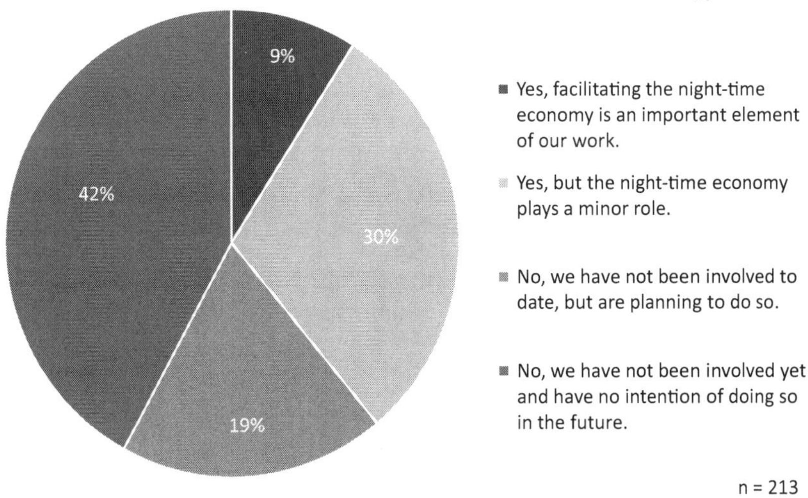

Figure 6.1 As a business development agency, are you already involved in the night-time economy? (n = 213).

Source: Konermann 2022, 148.

Table 6.1 Number of employees in the restaurant and bar industry in Germany (2019–2022, DEHOGA 2022, 5; DEHOGA 2023, 4–5; cutoff date 30.09).

	Employees 2019	*Employees 2020*	*Employees 2021*	*Employees 2022*
Restaurants (food-oriented)	1.266.383	1.165.433	1.086.557	1.160.243
Bars and Cafés (beverage-oriented)	227.544	148.714	155.587	199.925
Restaurant and bar industry	1.493.927	1.314.147	1.242.144	1.360.168

and companies. While this has led to corresponding attention to this economic sector in Great Britain and North America, it has been considerably less pronounced in Germany. However, even in Germany, the customers' decision for a place of residence is certainly influenced by the restaurants and bars on location (Lichtblau et al. 2017). Similar to the role of the restaurant and bar industry influencing the attractiveness of destinations for tourists (Kivela and Crotts 2006). The importance of local nightlife is increasingly being taken into account not only in tourism marketing but also in the context of location marketing with the target groups of companies and professionals, to convey the image of a lively and liveable city (Krüger, Schmid, and Jauernig 2015).

In addition to these rather soft factors, the restaurant and bar industry also has direct economic effects. Restaurants and bars create jobs and strengthen municipal tax revenues. Due to the medium-sized character of this industry (Lichtblau et al. 2017), the operators usually live in the local area of the respective business and thus contribute to the local economy (Franz 2020a). At the same time, restaurants are a relatively labor-intensive economic sector. In 2022, 1.3 million people were employed in the restaurant and bar industry in Germany (DEHOGA 2023, 4–5, see Table 6.1). Purchasing goods and services for restaurants and bars generates further economic benefits. Through coupling effects, restaurants and bars can also have economic significance for a location since customers who visit a restaurant often also shop in the nearby area or use other services around (Franz 2020a).

The restaurant and bar industry is, therefore, not only of direct economic importance but is also an asset in terms of enhancing the image and attractiveness of city districts and entire cities. With the declining attractiveness of the retail sector and the lack of skilled labor, economic development, and city marketing agencies should not neglect this sector and actively promote its development.

6.4 What Changes Already Occurred in the Restaurant and Bar Industry Before the COVID-19 Pandemic?

Since the 1970s the restaurant and bar industry in Germany has been undergoing significant changes, although the intensity and timing of these changes vary considerably depending on the region. A considerable decrease in the number of

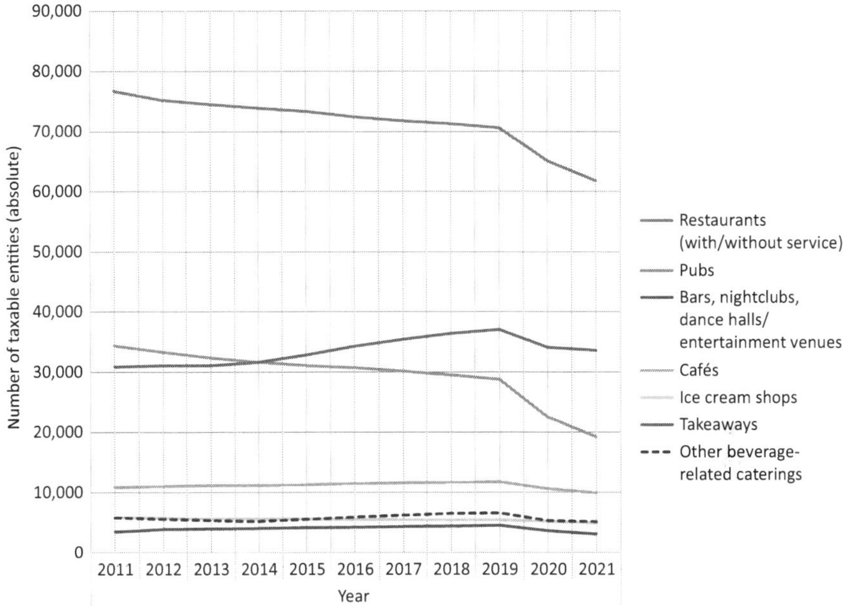

Figure 6.2 Number of restaurant and bar companies by type 2009–2021 (DEHOGA 2022, 12; DEHOGA 2023, 11).

restaurants and bars is accompanied by a change in format: there are fewer restaurants, bars, and pubs but more snack bars and cafés (see Figure 6.2). At the same time, there has been a concentration of restaurants and bars in prime locations for a long time, while closures are taking place in residential, suburban, and rural areas. Recently, it has been observed that more than before restaurants and bars are moving into city centers (Franz 2020b). Overall, the share of system caterers is increasing – especially in prime locations with high rents and as non-integrated free standers in commercial areas on main roads (Schäfer 2017).

A variety of reasons can be identified for this change depending on the context. First, there is a change in consumer demands in lifestyle and communication behavior, and in working and living conditions. These changes impact the type and frequency of the restaurant and bar services demanded. The guests' demands on the breadth, depth, and quality of the offer, as well as the space available and room design, have increased. This is also reflected in this quote from an interview with a bartender in 2018:

> In the past, you just had a beer or a long drink with a meatball or sausage, but nowadays [...] you have to offer the whole range. And then someone says to me: 'It used to be quicker here with the service'. I say: 'Yes, in the old days you didn't order a caffeine-free latte macchiato with caramel and you didn't want the prawn skewers with it either'.

These demand-side effects are compounded by the reduced bargaining power of restaurant and bar owners when dealing with breweries and wholesalers. On top, there is increased competition with clubhouses and large-scale system caterers. Another vital aspect is the increasing regulatory pressure and enforcement of rules and laws. One interviewee, referring to the situation in the city center, says, "In the last 20 years [...] the regulations for restaurants and bars have been as high as possible and even [...] [there are] statutes for outdoor catering" (interview with consultant 2023). But some factors are rooted in the restaurant and bar businesses themselves, for instance, the lack of innovation in some companies, staff shortages and the lack of succession, an investment backlog combined with low capitalization, and difficult access to investors. Ultimately, the standard of requirements an owner has to meet has increased. Consequently, it is becoming more difficult to successfully maintain owner-operated restaurants and bars (Franz 2020b).

Many companies were already in a difficult situation pre-pandemic. This is also stressed by the following quote from an interview with a restaurant owner in 2020: "Many in the restaurant business are just scraping by, that's really bad, they just have their income and that's it. And when the income stops coming in, then there's an issue [...]".

Due to this difficult situation, it can be assumed that many companies had a low level of resilience before the pandemic. Against this background, the slump in the restaurant and bar industry triggered by the COVID-19 crisis occurred in 2020.

6.5 How Has the COVID-19 Pandemic Affected the Restaurant and Bar Industry in Germany?

At the end of February 2020, there was a rapid increase in COVID-19 infections in Germany. On March 13, 2020, Chancellor Angela Merkel called on citizens to cancel all unnecessary events and refrain from socializing. At the same time, measures were taken in many federal states to contain the spread of COVID-19. These included a ban on large events, which led to decreased sales in bars and restaurants. On March 16, 2020, the federal government and the heads of state governments decided to close bars, clubs, discos, Kneipen (German pubs), and similar establishments to the public. In addition, measures were adopted for refectories, restaurants, and hotels to minimize the risk of spreading the coronavirus. These included regulations of table spacing, the number of guests, restriction of opening hours, hygiene measures, and instructions. Shortly thereafter, the regulations were further tightened after the renowned Robert Koch Institute classified the risk to public health as high on March 17. The individual states introduced further restrictions between March 17, 2020 and March 22, 2020. All restaurants and bars were closed for on-the-spot consumption, but take-away and delivery services remained allowed – with various hygiene requirements. While the restaurant and bar industry was reopened successively in different federal states in May 2020, various restrictions remained, which meant additional costs and lower turnover for businesses. Overall, some 41 percent of the surveyed companies compensate for parts of the revenue loss by offering takeaway and delivery services. In addition, consumers reacted to the continuing COVID-19 threat with restrained consumption at restaurants and bars. In May 2020, for example, sales in the restaurant and bar industry fell by an average of 53.6 percent compared to the same period of the previous year (Destatis 2020). From November 2 onwards, there was a renewed closure of restaurants and bars, which was only gradually lifted in May 2021. Overall, turnover in the restaurant and bar industry in Germany fell by around 14.2 billion euros in 2020 compared to 2019, a drop of almost 27.5 percent (see Figure 6.3, DEHOGA 2021, n.p.). This is also reflected in the development of employee numbers. In 2019, 1.5 million people were employed in the restaurant and bar industry; in 2020 it was still 1.3 million people, with jobs being lost, especially in the beverage-dominated pub and bar industry. Compared with 2019, in 2020 a third fewer employees were working in pubs and bars (cut-off date 30.9, DEHOGA 2022, 5). This shows that these areas were particularly affected by the pandemic.

Only very few businesses were financially prepared for such collapses, as was shown in the interview with a restaurant owner in 2020: "We didn't expect anything like this. You always reckon that there will be some kind of slump when people are financially worse off [...], but you didn't expect such a serious slump [...]".

Nearly all the businesses we surveyed have suffered significant financial problems as a result of the Corona crisis. Various support measures and loans, often in combination, were used extensively to alleviate the financial difficulties. Loans from the federal government and the federal states so-called Coronahilfen (Corona aids) were used by 85 percent of the enterprises surveyed.

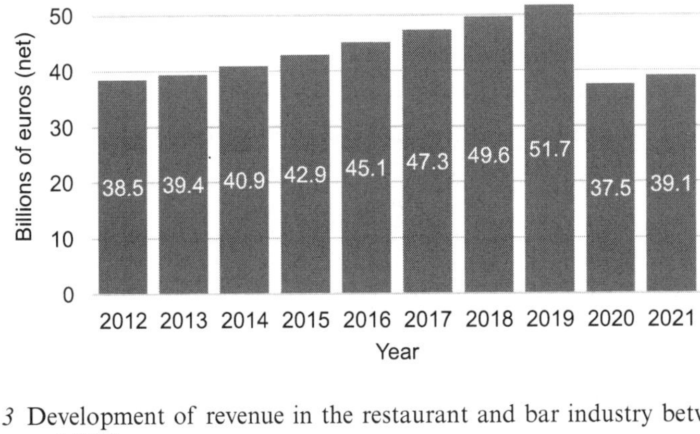

Figure 6.3 Development of revenue in the restaurant and bar industry between 2011 and 2021.

Source: Destatis 2023, n.p.

The Kurzarbeitergeld (short-time work compensation) was also actively used by restaurateurs and bar owners (77.9 percent). Three-quarters (75.4 percent) of the businesses used equity capital and operating profits, which, however, were quickly used up as the loss of income continued. Many restaurateurs and bar owners have little or no reserves, so the massive drop in revenue has also had a strong impact on the private sector. Our survey showed that about 35 percent of the operators had to limit their private finances. For only 11 percent did the crisis have little or no impact on the private sector. The companies surveyed were mostly pessimistic about the future. They also expected lower profits in the long term, increased hygiene regulations, a lack of capital for investments, and fewer customers. Nearly 55 percent of the surveyed companies perceived their business's future as precarious. Among companies that had already experienced declining turnover, the figure was 76 percent. Owner-managed businesses are much more at risk than non-owner-managed businesses.

> [I] am a family man and have been going through this for three years [...] and of course, at some point, this also wears you out, and at some point, you go to pieces because you keep telling yourself that this is coming [to get better off financially] and then it does not come again. Once you get on your face, you get up again, [and then] you get on your face again – no support from politics, no support [...], no relief on the side.
>
> (interview with restaurant owner 2023)

Based on these results, it is to be expected that the structural and format change in the restaurant and bar industry will be considerably intensified.

The experience from 2020 has shown that the affected companies use various short-term coping measures. As Biggs, Hall, and Stoeckl (2012) have noted, such measures can range from cost reductions to marketing activities to (temporary) closures. However, some of the measures that are helpful in the short term may have unintended long-term consequences for business resilience. This is illustrated by the examples of staff layoffs and the use of state loans during the crisis.

> In July or August 2021, the kitchen [employees] approached me, and shortly before the end of their probationary period, they thought, 'Now the restaurant and bar industry will get going again.' And then I found out that they had only used this [job] as a stopover. Then the entire kitchen quit. [...] And then I had to close the [one of two] restaurant[s]. Everyone was laid off, complete layoff here, the service team was out the door, and the kitchen was finally closed.
>
> (interview with restaurant owner 2023)

For example, the layoffs during the closures led to staff shortages in many restaurant and bar businesses after they reopened, and the taking out of loans during the crisis enabled solvency but subsequently put a permanent strain on the financial situation of the businesses.

Our survey during the first COVID-19 wave showed that 41 percent of the businesses surveyed tried to compensate for the loss of sales by offering delivery or takeaway of their goods. Over 30 percent have been introduced due to the COVID-19 pandemic, with takeaway offers predominating. Businesses that offered takeaway/delivery services suffered a smaller decline in revenue (82.6 percent) than those that did not (95.4 percent; see Figure 6.4), although this does not mean the service was profitable. In rural areas, the share of businesses (36.8 percent) with takeaway and delivery services was surprisingly similar to that in cities (40.1 percent). At the same time, however, the respondents also stated that delivery services, particularly, were hardly profitable in rural areas because the distances were too long and demand was low. Consequently, takeaway at weekends or on public holidays was mainly offered. At the same time, the majority of the surveyed businesses did not offer takeaway or delivery services during the first COVID-19 wave, as the effort to change the offer was too great:

> If you do it professionally, you need special dishes where you can seal them because they take them with them or so. [...] That's why I thought, if we're only closed for four weeks, I won't start. [...] Because I didn't really feel like investing in the middle of it [the COVID-19 pandemic].
>
> (Interview with restaurant owner 2020)

Especially for the beverage-oriented bar industry, takeaway, and delivery services were difficult or impossible to implement due to their products. Where an introduction occurred, the short-term coping measure of introducing delivery service and/or takeaway options could also represent a permanent strategic realignment of operations.

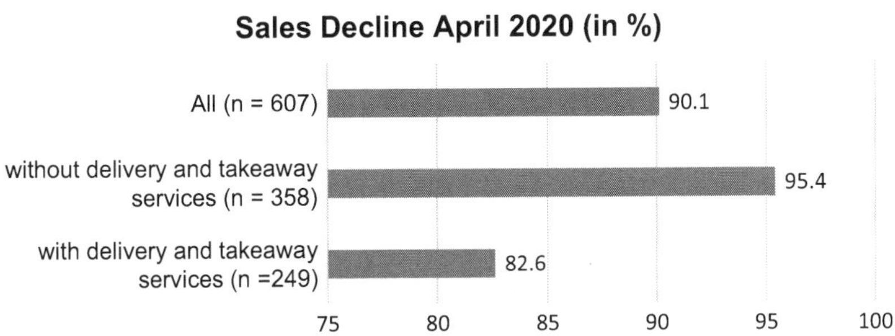

Figure 6.4 Effect of delivery and takeaway services on sales decline in April 2020 in percent (data: own survey).

6.6 What Strategies Can Business Development and City Marketing Agencies Employ to Enhance the Resilience of the Local Restaurant and Bar Industry During Crisis Situations?

During the COVID-19 pandemic, many municipalities have taken various measures to strengthen the short-term coping capacity of their restaurants and bars. These include municipalities suspending or waiving public fees, charges, and taxes to help companies. This is especially true for special use fees for outdoor catering – examples include Bremen, Dresden, and Lübeck. Business development agencies have also provided information for companies. Especially local restaurant and bar owners with small businesses as well as those with a migration background, are often not members of DEHOGA and lack information on support possibilities. This is where business development agencies can help by providing information. Another support possibility is the expansion of outdoor catering – for example, through the (temporary) rededication of parking spaces to improve income opportunities. This option was implemented by many municipalities in the short term during the pandemic – for example, in Mannheim, where there were successful field trials in this regard even before COVID-19. Extending the permitted opening hours for outdoor catering could also provide additional revenue for establishments. However, as the latter was often seen as conflicting with the pandemic measures, there were hardly any extensions to the permitted opening hours during the pandemic period (on the contrary: in some cases, the permitted opening hours were shortened to contain contacts). In addition to such short-term measures, municipalities should first develop a restaurant and bar industry strategy or concept to strengthen the resilience of the sector (Jones et al. 2003). While many German municipalities already have concepts for the retail sector, there are no such concepts for the restaurant and bar industry in Germany. However, individual municipalities have developed strategies to promote the restaurant and bar industry and/or nightlife (e.g. Mannheim, Schmid 2018) or have included corresponding aspects in concepts for the development of their city centers or certain other neighborhoods (e.g. the city center concept Hannover City 2020, see Krüger, Schmid, and Jauernig 2015). Further examples of the restaurant and bar industry or night economy concepts and strategies can be found abroad. For example, the city of Bern adopted the concept of Nightlife Bern as early as 2013, and the city of Zurich declared nightlife a strategic focus in 2016 (Schmid 2018).

In general, municipalities should strengthen expertise on the problems and needs of the restaurant and bar industry in their relevant departments and promote knowledge exchange (Krüger, Schmid, and Jauernig 2015). During the COVID-19 pandemic, it became apparent that many municipalities lacked the necessary know-how to develop the abruptly required support measures for the restaurant and bar industry. "A lot of [municipal actors] [...] think, 'yeah, restaurants and bars, [...] anybody can do it, then I can talk about it too'. And usually, nothing comes out of it, or [...] the wrong restaurateurs [are attracted]"

(Interview with consultant 2023). Within the administration, clear responsibilities for the restaurant industry should be designated. This applies both to the clarification of issues within the administration, also to avoid conflicts of objectives (Krüger, Schmid, and Jauernig 2015), as well as to the existence of uniform contact persons for restaurants and bars. It is quite problematic that in most municipalities, restaurant and bar operators are confronted with constantly changing contact persons and a wide variety of responsibilities. "In many cases, they [the entrepreneurs] have not been able to reach a contact person [...] they [...] have not reached anyone at all, not even by phone or by emails" (interview with economic development agency 2023). This makes communication and the development of the sector more difficult. Especially in crisis situations, this has the disadvantages of slowing down communication and increasing transaction costs for companies.

To strengthen communication with the night economy sectors and their representation in a municipality, more and more municipalities are introducing the position of night mayor. In 2018, Mannheim was the first city in Germany to establish such an interface between nightlife actors and stakeholders and the city administration. The city of Stuttgart 2021 has taken a special approach: in addition to the position of night mayor (which is called night manager there), the position of the project manager of the nightlife coordination office has been established. This tandem solution aims to ensure that an interface is formed between private and public interests, combining insider knowledge from both sides and, at the same time, providing a direct link to the administration.

As argued above, formal and informal relationships with actors outside the company can ensure important access to resources (Stevenson 2014) and thus strengthen the resilience of businesses. In addition to material resources, access to intangible resources such as social capital, knowledge, and information is vital to help businesses cope with crises in the short term and to develop adaptation strategies. Especially networking is often low in owner-operated restaurant and bar establishments. For example, a much smaller proportion of them are organized in advertising or location associations than is the case with retail businesses. Many restaurant owners are members of DEHOGA. However, DEHOGA rarely represents their interests at the local level and can seldom provide local networking effects (interviews with restaurant owners and DEHOGA representatives). To strengthen the network relations and the knowledge transfer among restaurants and bars, municipalities or their institutions for business development and city marketing should create regular forums for the exchange between restaurant and bar owners. Possible formats are, for example, regulars' tables or lecture events in which current developments and innovations are presented. Such formats can serve, on the one hand, to increase the transfer of knowledge and communication within the industry, and on the other hand, to increase communication between business development and city marketing agencies as well as restaurant and bar owners. This can also help to build trust and reduce conflicts. If there are agglomerations of restaurants and/or bars, establishing formalized location communities should be considered. One possible approach is the concept of Business Improvement Districts, as applied in Bochum, for example, for the entertainment district known as the Bermudadreieck (Bermuda Triangle). Business Improvement Districts offer the possibility of financing measures, improving the management of the restaurant and bar agglomeration, and strengthening communication between the different actors (Franz 2020a).

Offering low-threshold training can also be useful. One possible topic is digitalization. The handling of the COVID-19 crisis has increased the trend toward digitalization in the restaurant and bar industry. As a result, digital solution providers and delivery service providers will increasingly skim off revenues, which will particularly disadvantage businesses with a low-price margin. This can also lead to new dependencies. Businesses should be supported in this regard with training and the organization of the exchange of experiences.

The organization of events such as festivals or city (sub)festivals can not only contribute to changing the image of the respective locations but also increase acceptance among the resident population (Krüger, Schmid, and Jauernig 2015, for a critical perspective, see Redepenning, Rhein, and Sauerwald 2016) but is also suitable for strengthening the income base of local restaurants. The problem is that up to now, it has often been mainly large restaurant and bar industries or beverage retailers that have been involved. To promote a diverse offer and strengthen the resilience of local businesses, as many different local restaurant and bar businesses as possible should be involved. In addition, restaurants should be supported in the organization of business succession. In many municipalities, economic development institutions support entrepreneurs in their search for business succession – but so far, corresponding offers do not always apply to restaurant and bar businesses. They should be extended to them (Franz 2020a).

6.7 Conclusion

The COVID-19 pandemic had a devastating impact on the restaurant and bar industry. Although the affected businesses have responded to the crisis with various coping measures and adaptation strategies, most companies were pessimistic about the future during the COVID-19 pandemic. Many of these businesses already had a low level of resilience before the pandemic, i.e., a low ability to react to crisis situations and to develop adaptation strategies. The reasons for the often difficult situation of restaurants and bars even before the pandemic are manifold. On the one hand, there are changes on the demand side: the extensive change in lifestyle and communication behavior as well as in working and living conditions, had a clear impact on the type and frequency of the food and drink services demanded. At the same time, the guests' demands on the breadth, depth, and quality of the offer, as well as the space and room design, had increased. In addition to these demand-side effects, there were regulatory influences as well as sectoral and, in part, company-internal factors. Significant here is the diminished bargaining power of restaurant and bar owners when dealing with breweries and wholesalers, the low capitalization with difficult access to credit, the increased competition, the increasing regulation and enforcement of rules and laws, the lack of innovation in some businesses, staff shortages and the lack of business succession. Against this background, the pandemic acted like an accelerant. The revenue of the businesses collapsed massively, and without state aid, hardly any restaurant and bar would have survived.

The need for action in many municipalities to strengthen the resilience of restaurant and bar businesses often only became clear during the COVID-19 pandemic, while the rather gradual structural and format changes that have been taking place for a long time in the industry were hardly noticed before. In addition to federal and state relief measures, many municipalities also took action during the COVID-19 pandemic to strengthen the short-term coping capacity of their restaurant and bar industry. This included, for example, suspending fees and charges, providing information, and allowing outdoor catering to expand. Even though the acute threat of the pandemic is over, municipal business development and city marketing institutions should not lose focus on the restaurant and bar industry again. The importance of the restaurant and bar industry for the (economic) development of cities – which has been underestimated for decades – not only results from the direct economic effects but also acts as a factor for the image and attractiveness of city districts and entire cities. Particularly against the background of the declining attraction of retail trade for city centers and the increasing importance of the attractiveness of cities for attracting sufficient skilled workers, business development and city marketing should no longer neglect this and contribute to making the restaurant and bar industry more resilient to future crises.

To strengthen the resilience of businesses in the restaurant and bar industry, a local restaurant and bar industry strategy or concept should be developed – similar to the retail or market and center concepts that exist in many German towns, and which specify where, and what type of retail can be located. Alternatively, attention should be paid to the restaurant and bar industry in center or retail concepts. The development of suitable location conditions for restaurants and bars can contribute to the company's resilience in the long term. In general, it is necessary to strengthen the restaurant and bar industry's expertise in business development and city marketing agencies and to designate clear responsibilities for this sector. One possible approach is appointing a night mayor, who should, consider the restaurant and bar industry as a whole. The approach of a dual leadership of persons with a restaurant and bar industry background and administrative background seems to make sense. In this way, communication barriers can be reduced, and competencies can be built up that allow faster and more competent reactions in crisis situations and contribute to strengthening the resilience of the restaurant and bar industry in the long term. In addition, formal and informal relationships with actors outside the company can ensure important access to resources. Especially in owner-operated restaurant and bar businesses, networking is often low, which leads to poorer access to knowledge and resources. In order to increase networking and knowledge transfer between businesses, but also communication between business development/city marketing and restaurant and bar owners, municipalities should create forums for exchange between operators of restaurants and bars. A good knowledge transfer can strengthen the adaptability and innovative capacity and, thus, ultimately, the profitability and resilience of businesses. In addition, restaurants and bars should be supported in the organization of business succession to prevent profitable and resilient businesses from closing down and thus unnecessarily weakening the restaurant and bar offer in the respective municipality.

As of now, the COVID-19 pandemic has officially ended, but its effects on the restaurant and bar industry are still prevalent. Overall, the sector is showing signs of recovery, with the number of employees rising significantly again in 2022 and there are even more jobs than in 2019 – i.e., before the Covid-19 pandemic (DEHOGA 2023, 4–5; see Table 6.1). In fact, system catering is recovering better from the pandemic than owner-operated restaurants. System catering accounted for 47 percent of total visits in 2022. This is 17 percent more visits than in 2021 but still represents a decrease of 4 percent compared to 2019. With a year-on-year increase in visits of 27 percent, individual catering accounted for 53 percent of the total market. Compared to 2019, however, the number of visits is still down 24 percent (Bundesverband der Systemgastronomie 2023, n.p.).

Many establishments are unsure of the long-term impact of the crisis on their business survival. Furthermore, the industry is currently facing other crises, such as staff shortages and high energy costs, which are causing further problems in Germany. It is crucial to conduct research on this subject, including the role of formal and informal network relations regarding the resilience of businesses within this industry. Additionally, policymakers and supporting institutions need to be more aware of and attuned to the concerns of the restaurant and bar sector.

Acknowledgments

We would like to thank the Niedersächsisches Ministerium für Wissenschaft und Kultur, which has supported the research with funds from "SPRUNG".

References

Alonso-Almeida, M. M., and K. Bremser. 2013. Strategic responses of the Spanish hospitality sector to the financial crisis. *International Journal of Hospitality Management* 32 (1):141–48. doi: 10.1016/j.ijhm.2012.05.004

Andres, L., and J. Round. 2015. The creative economy in a context of transition: A review of the mechanisms of micro-resilience. *Cities* 45:1–6. doi: 10.1016/j.cities.2015.02.003

Baier R., and K. Engelen. 2015. Straßenumgestaltungen nach dem sogenannten "Shared Space"-Gedanken–/empirische Befunde aus Deutschland und der Schweiz [Street redesigns based on the so-called "shared space" concept/empirical findings from Germany and Switzerland]. *Straßenverkehrstechnik* 5 (2015):317–22.

Berkhout, F. 2012. Adaptation to climate change by organizations. *Wiley Interdisciplinary Reviews: Climate Change* 3 (1):91–106. doi: 10.1002/wcc.154

Berman, R., C. Quinn, and J. Paavola. 2012. The role of institutions in the transformation of coping capacity to sustainable adaptive capacity. *Environmental Development* 2:86–100. doi: 10.1016/j.envdev.2012.03.017

Biggs, D., C. M. Hall, and N. Stoeckl. 2012. The resilience of formal and informal tourism enterprises to disasters: reef tourism in Phuket, Thailand. *Journal of Sustainable Tourism* 20 (5):645–65. doi: 10.1080/09669582.2011.630080

Billington, M. G., J. Karlsen, L. Mathisen, and I. B. Pettersen. 2017. Unfolding the relationship between resilient firms and the region. *European Planning Studies* 25:425–42. doi: 10.1080/09654313.2016.1276886

Boschma, R. 2015. Towards an Evolutionary Perspective on Regional Resilience. *Regional Studies* 49 (5):733–51. doi: 10.1080/00343404.2014.959481

Bristow, G., and A. Healy. 2014. Regional Resilience: An Agency Perspective. *Regional Studies* 48 (5):923–35. doi: 10.1080/00343404.2013.854879

Brizek, M. G., R. E. Frash, B. M. McLeod, and M. O. Patience. 2021. Independent restaurant operator perspectives in the wake of the COVID-19 pandemic. *International Journal of Hospitality Management* 93:102766. doi: 10.1016/j.ijhm.2020.102766

Bundesverband der Systemgastronomie. 2023. Branchendaten [sector data]. Accessed March 13, 2024. www.bundesverband-systemgastronomie.de/de/politik/branchendaten.html

Crîşmariu, O. D., M. O. Tănase, I. Tenie, and A. Abdallah. 2022. The response of the Romanian hospitality industry to the COVID-19 pandemic. Focus on the restaurant industry. *Cactus Tourism Journal* 4 (1):87–94. doi: 10.24818/CTS/4/2022/1.08

Conz, E., and G. Magnani. 2019. A dynamic perspective on the resilience of firms: A systematic literature review and a framework for future research. *European Management Journal* 38 (3):400–12. doi: 10.1016/j.emj.2019.12.004

DEHOGA (Deutscher Hotel- und Gaststättenverband). 2022. DEHOGA Zahlenspiegel IV/2022 [DEHOGA statistical review IV/2022]. Accessed January 27, 2024. https://www.dehoga-bundesverband.de/fileadmin/Startseite/04_Zahlen___Fakten/07_Zahlenspiegel___Branchenberichte/Zahlenspiegel/DEHOGA-Zahlenspiegel_4._Quartal_2021.pdf

DEHOGA (Deutscher Hotel- und Gaststättenverband). 2023. DEHOGA Zahlenspiegel I/2023 [DEHOGA statistical review I/2023]. Accessed January 27, 2024. https://www.dehoga-bundesverband.de/fileadmin/Startseite/04_Zahlen___Fakten/07_Zahlenspiegel___Branchenberichte/Zahlenspiegel/DEHOGA-Zahlenspiegel_1._Quartal_2023.pdf

DEHOGA (Deutscher Hotel- und Gaststättenverband). 2021. DEHOGA Zahlenspiegel II/2021. [DEHOGA statistical review II/2021]. Accessed July 21, 2023. https://www.dehoga-bundesverband.de/fileadmin/Startseite/04_Zahlen___Fakten/07_Zahlenspiegel___Branchenberichte/Zahlenspiegel/DEHOGA-Zahlenspiegel_2._Quartal_2021.pdf

Destatis (2023). Umsatzsteuerpflichtige, Steuerbarer Umsatz, Umsatzstezer (Voranmeldung): Deutschland, Jahr, Wirtschaftszweige [Taxable persons, taxable turnover, turnover tax (advance return): Germany, year, economic sectors]. Accessed January 27, 2024. https://www-genesis.destatis.de/genesis/online?operation=find&suchanweisung_language=de&query=73311-0002#abreadcrumb

Destatis. 2020. Gastgewerbeumsatz im Mai 2020 um 45 % gegenüber Vormonat gestiegen. Pressemitteilung Nr. 272 vom 20. Juli 2020 [Hospitality industry revenue up 45% in May 2020 compared to previous month. Press release no. 272 from 20 July 2020]. Accessed March 4, 2022. https://www.destatis.de/DE/Presse/Pressemitteilungen/2020/07/PD20_272_45213.html

Esteve-Pérez, S., and J. A. Mañez-Castillejo. 2008. The resource-based theory of the firm and firm survival. *Small Business Economics* 30 (3):231–49. doi: 10.1007/s11187-006-9011-4

Florida, R. 2005. *Cities and the creative class*. New York: Routledge.

Franz, M. 2020a. Gaststätten als Teil der lokalen Ökonomie [Restaurants as part of the local economy]. In *Lokale Ökonomie–Konzepte, Quartierskontexte und Interventionen* [Local economy concepts, neighbourhood contexts and interventions], ed. S. Henn, M. Behling, and S. Schäfer, 191–209. Berlin: Springer Spektrum. doi: 10.1007/978-3-662-57780-6_14

Franz, M. 2020b. Treffpunkt Gaststätte – warum sich Kommunen stärker um ihre Kneipen und Restaurants kümmern sollten [Meeting place restaurant - why municipalities should take more care of their pubs and restaurants]. *Standort – Zeitschrift für angewandte Geographie* 44:93–98. doi: 10.1007/s00548-020-00635-0

Franz, M., P. Müller, and T. Neise. 2020. Im Krisenmodus – Die Auswirkungen der Corona-Pandemie auf die Gastronomie in Deutschland [In crisis mode - The impact of the COVID-19 pandemic on the gastronomy sector in Germany]. *Berichte des Arbeitskreises Geographische Handelsforschung* 47:11–12. www.geographie.hu-berlin.de/de/abteilungen/wirtschaftsgeographie/AK-Einzelhandelsforschung/publikationen/berichte/berichte-des-arbeitskreises-geographische-handelsforschung-nr.47/at_download/file

Gittell, J. H., K. Cameron, S. Lim, and V. Rivas. 2006. Relationships, layoffs, and organizational resilience: Airline industry responses to September 11. *The Journal of Applied Behavioral Science* 42 (3):300–329. doi: 10.1177/0021886306286466

Hakim, M. P., L. D. A. Zanetta, and D. T. da Cunha. 2021. Should I stay, or should I go? Consumers' perceived risk and intention to visit restaurants during the COVID-19 pandemic in Brazil. *Food Research International* 141:110152. doi: 10.1016/j.foodres.2021.110152

Headd, B. 2003. Redefining business success: Distinguishing between closure and failure. *Small Business Economics* 21:51–61. doi: 10.1023/A:1024433630958

Jones, P., A. Charlesworth, V. Simms, D. Hillier, and D. Comfort. 2003. The management challenges of the evening and late night economy within town and city centres. *Management Research News* 26 (10/11):96–104. doi: 10.1108/01409170310784096

Kaniovski, S., and M. Peneder. 2008. Determinants of firm survival: a duration analysis using the generalized gamma distribution. *Empirica* 35 (1):41–58. doi: 10.1007/s10663-007-9050-3

Khan, M. A. U., and M. A. Sayem. 2013. Understanding recovery of small enterprises from natural disaster. *Environmental Hazards* 12 (3-4):218–39. doi: 10.1080/17477891.2012.761593

Kivela, J., and J. C. Crotts. 2006. Tourism and gastronomy: Gastronomy's influence on how tourists experience a destination. *Journal of Hospitality & Tourism Research* 30 (3):354–77. doi: 10.1177/1096348006286797

Konermann, J. 2022. Nachtökonomie als Themenfeld der kommunalen Wirtschaftsförderung–Status quo, Herausforderungen und Förderansätze [Night-time economy as a topic area of municipal economic development - status quo, challenges and approaches to funding]. *Standort Zeitschrift für angewandte Geographie* 145–50. doi: 10.1007/s00548-022-00816-z

Krüger, T., J. F. Schmid, and T. Jauernig 2015. Stadtnachacht – Management der Urbanen Nachtökonomie [City after eight – Management of the Urban Night Economy]. Accessed April 6, 2017. www.hcu-hamburg.de/fileadmin/documents/Professoren_und_Mitarbeiter/Projektentwicklung__-management/Forschung/stadtnachacht_2015.pdf

Le, N. P. 2015. Small Ethnic Restaurant Business Owner Strategies to Remain Operable Beyond 7 Years. Doctoral diss., Walden University.

Lengnick-Hall, C. A., T. E. Beck, M. L. Lengnick-Hall. 2011. Developing a capacity for organizational resilience through strategic human resource management. *Human Resource Management Review* 21 (3):243–55. doi: 10.1016/j.hrmr.2010.07.001

Li, B., Y. Y. Zhong, T. Zhang, and N. Hua. 2021. Transcending the COVID-19 crisis: Business resilience and innovation of the restaurant industry in China. *Journal of Hospitality and Tourism Management* 49:44–53. doi: 10.1016/j.jhtm.2021.08.024

Lichtblau, K., R. Bertenrath, A. Kleissner, H. Kempermann, A. Millack, and J. Ewald. 2017. *Die Bedeutung des Hotel- und Gaststättengewerbes [The importance of the hospitality industry]*. Köln: Institut der deutschen Wirtschaft Köln Consult GmbH.

Linnenluecke, M. K., and A. Griffiths. 2015. *The climate resilient organization. Adaptation and resilience to climate change and weather extremes*. Cheltenham: Edward Elgar Publishing.

Marcos, J., and S. Macaulay. 2008. *Organisational Resilience: The Key to Anticipation, Adaptation and Recovery*. Cranfield: Cranfield School of Management.

Martin, R. L. 2018. Shocking aspects of regional development: towards an economic geography of resilience. In *The New Oxford Handbook of Economic Geography*, ed. G. Clark, M. Gertler, M. P. Feldman, and D. Wójcik, 839–864. Oxford: Oxford University Press. doi: 10.1093/oxfordhb/9780198755609.013.43

Matsenko, O., O. Kubatko, V. Bardachenko, and K. Demchuk. 2021. Transformation of the Restaurant Business as a Result of the COVID-19 Pandemic: Improving the Security of Service and Maintaining the Health of Human Capital. *Health Economics and Management Review* 3:27–38. doi: 10.21272/hem.2021.3-03

Mayring, P. 2014. Qualitative content analysis: theoretical foundation, basic procedures and software solution. Klagenfurt. https://nbn-resolving.org/urn:nbn:de:0168-ssoar-395173 Accessed July 16, 2023.

Mitrofanova, I. V., O. A. Chernova, and V. V. Batmanova. 2022. Digitalization of business processes in adaptation of catering industry to new realities: Covid-19 pandemic. *Serbian Journal of Management* 17 (1):237–51. doi: 10.5937/sjm17-34603

Neise, T., J. Revilla Diez, and M. Garschagen. 2018. Firms as drivers of integrative adaptive regional development in the context of environmental hazards in developing countries and emerging economies – A conceptual framework. *Environment and Planning C: Politics and Space* 36 (8):1522–41. doi: 10.1177/2399654418771079

Neise, T., and J. Revilla Diez. 2019. Adapt, move or surrender? Manufacturing firms' routines and dynamic capabilities on flood risk reduction in coastal cities of Indonesia. *International Journal of Disaster Risk Reduction* 33:332–42. doi: 10.1016/j.ijdrr.2018.10.018

Neise, T., P. Verfürth, and M. Franz. 2021. Rapid responding to the COVID-19 crisis: Assessing the resilience in the German restaurant and bar industry. *International Journal of Hospitality Management* 96:102960. doi: 10.1016/j.ijhm.2021.102960

Parsa, H. G., J. T. Self, D. Njite, and T. King. 2005. Why restaurants fail. *Cornell Hotel and Restaurant Administration Quarterly* 46(3):304–22. doi: 10.1177/0010880405275598

Parsa, H. G., J. Self, S. Sydnor-Busso, H. J. Yoon. 2011. Why Restaurants Fail? Part II - The Impact of Affiliation, Location, and Size on Restaurant Failures: Results from a Survival Analysis. *Journal of Foodservice Business Research* 14 (4):360–79. doi: 10.1080/15378020.2011.625824

Parsa, H. G., J. P. I. van der Rest, S. R. Smith, R. A. Parsa, and M. Bujisic. 2015. Why restaurants fail? Part IV: The relationship between restaurant failures and demographic factors. *Cornell Hospitality Quarterly* 56 (1):80–90. doi: 10.1177/1938965514551959

Pätzold, R. 2012. Tischlein, deck dich! Über städtische Gastronomielandschaften und was Stadtplanung damit zu tun haben könnte [Table-be-set! On urban gastronomy landscapes and what urban planning might have to do with them]. In *Proceedings Real Corp 2012*, ed. M. Schrenk, V. V. Popovich, P. Zeile, and P. Elisei. http://geomultimedia.org/archive/CORP2012_132.pdf Accessed September 23, 2016.

Pätzold, R. 2014. Sind Gastronomen (auch) Stadtentwickler? [Are gastronomes (also) urban developers?]. Standort – *Zeitschrift für angewandte Geographie* 38 (4):225–29. doi: 10.1007/s00548-014-0351-z

Pohl, T. 2010. Reurbanisierung als Trend? Folgen des Wandels der raumzeitlichen Alltagsorganisation in Familienhaushalten [Reurbanisation as a trend? Consequences of the change in the spatio-temporal organisation of everyday life in family households]. In *Zeiten und Räume der Stadt: Theorie und Praxis* [Times and spaces of the city: theory and practice], ed. D. Läpple, U. Mückenberger, and J. Oßenbrügge, 45–64. Opladen: Verlag Barbara Budrich.

Redepenning, M., N. Rhein, and D. Sauerwald. 2016. Eventorientierte Veranstaltungen und Lebensqualität innenstädtischer Wohnbevölkerung [Event-orientated activities and quality of life for inner-city residents]. *Standort – Zeitschrift für angewandte Geographie* 40 (3):170–76. doi: 10.1007/s00548-016-0439-8

Schäfer, L. 2017. Nicht-integrierte Standorte in der Fullservicegastronomie – Ansiedlungsmuster in Deutschland [Non-integrated locations in full-service gastronomy - settlement patterns in Germany]. Unpublished Bachelor Thesis, Geographical Institute, University of Osnabrück.

Schmid, J. F. 2018. Stadtnachacht – Stadtentwicklungspolitische Instrumente für das Management der urbanen Nachtökonomie [City after eight - Urban development policy instruments for the management of the urban night-time economy]. Dissertation, HafenCity University Hamburg.

Schneider, C. 2008. *Erfolgsfaktoren in kleinen Dienstleistungsunternehmen: eine Analyse am Beispiel der Gastronomie* [Success factors in small service companies: an analysis using the example of the gastronomy sector]. Wiesbaden: Gabler.

Scuderi, R., G. Tesoriere, and V. Fasone. 2020. Does Location Matter for Micro Shops Resilience? Evidence from Uganda. *International Regional Science Review* 92 (2):016001762093158. doi: 10.1177/0160017620931582

Stepper, M. 2015. Einkaufsstandort Innenstadt. Qualifizierung innerstädtischer Einzelhandelslagen vor dem Hintergrund des zunehmenden Online-Shopping [City center shopping location. Qualification of city center retail locations against the backdrop of increasing online shopping]. Dissertation, Technische Universität Kaiserslautern.

Stevenson, J. 2014. Organisational resilience after the Canterbury earthquakes: a contextual approach. Dissertation, University of Canterbury.

Tibay, V., J. Miller, A. Y. Chang-Richards, T. Egbelakin, E. Seville, and S. Wilkinson. 2018. Business resilience: A study of Auckland hospitality sector. *Procedia Engineering* 212:1217–24. doi: 10.1016/j.proeng.2018.01.157

Tsiapa, M., and I. Batsiolas. 2018. Firm Resilience in Regions of Eastern Europe during the Period 2007–2011. *Post-Communist Economies* 31 (1):19–35. doi: 10.1080/14631377.2018.1443250

Verfürth, P., T. Neise, M. Franz, and F. Sohns. 2022. The uneven effects of COVID-19 on the German restaurant and bar industry. *Erdkunde* 76 (2):127–140. doi: 10.3112/erdkunde.2022.02.05

7 Toward a Green Transformation Based on Eco-Innovations of Enterprises in Old Industrial Areas

The Case of Silesia Province

Michał Męczyński and Przemysław Ciesiółka

7.1 Introduction

Since the mid-1980s, there has been an ongoing interest in regional studies focusing on challenges and strategies regarding the transformation of old industrial areas (Boschma and Lambooy 1999; Hassink 1993; Trippl and Otto 2009). This literature is primarily geared to identifying and analyzing typical problems related to adopting innovations and focusing on path dependence and lock-in issues which lead to crisis. Coenen, Moodysson, and Martin (2015, 853) state that old industrial areas are typically considered as regions overspecialized in mature technologies and industries that cause their decline, thus facing lock-in. The dominance of mature sectors based on fossil fuels in these regions and a lack of entrepreneurship activity cause uncertainty and risk related to the transformation. Therefore, innovation activities in these regions often follow mature technological trajectories mainly of an incremental character, which, particularly in these areas, is associated with high risk and uncertainty (Grabher 1993; Tödtling 1992). Efforts to introduce radically new products into the market tend to be limited; this course of action causes lock-in and path-dependency, which hinders restructuring processes and the green transformation (Coenen, Moodysson, and Martin 2015; Hassink and Shin 2005; Jakobsen et al. 2022).

Well-known examples of old industrial regions locked in because of overspecialization in heavy industries are the Ruhr area in Germany (Grabher 1993), North East England (Coenen 2007; Hudson 1994), Wales (Morgan 2012) or southern Poland – Silesia Province (Gwosdz 2012, 2014). In these regions, the socio-economic transformation triggered by the collapse of old branches of industry and the deteriorating economic situation occurred at different rates. Owing to the centrally planned economy operating in Poland until 1989, the transformations in Silesia Province were the latest to start. However, despite restructuring programs implemented in this region for over thirty years, it is well-known for having severe difficulties that cause a lock-in related to traditional and fossil fuel-based industries and the concentration of environmental waste and pollution (such as carbon emissions, e.g. Gwosdz 2014; Tkocz 2014; Wiedermann 2007). Because of the traditional industry heritage,

DOI: 10.4324/9781003457626-9

the region has been dealing with crisis for decades. However, as Gwosdz (2014) indicated, the unfavorable situation in the old industrial region does not mean that technological changes and improvement in the quality of the natural environment cannot occur at all. Friedman (1992) states that the changes may be often the consequences of a crisis which triggers industrial, organizational, and institutional renewal processes and reinforcement of innovations. Conversely, Henderson (2015) claims that the changes may also be caused by external interventions in the form of a multi-level regional policy, either European, national, or regional (OECD 2019). An example of such a policy that can be a trigger for a renewal of Silesia Province is the European Commission European Green Deal (EGD) introduced in 2019 and followed by the new Circular Economy Action Plan (CEAP) introduced in 2020 as an improved 2015 plan. They aim to reduce pollution and the pressure on natural resources and create sustainable growth. The EGD and CEAP trigger industrial, organizational, and institutional renewal processes, reinforcing the implementation of eco-innovations (EIs) in enterprises. Defined as those that reduce the environmental impact of consumption and production activities, EIs play a very relevant role in the quest for more competitive, environmentally and socially sustainable societies (Carrillo-Hermosilla, Del Río, and Könnölä 2010; Machiba 2010). The creation and development of EIs in enterprises can play a vital role in the green transition of old industrial regions and their final escape from a negative lock-in and crisis. Based on the research carried out by Coenen, Moodysson, and Martin (2015) and Martin and Sunley (2006), the adoption process of innovations in old industrial regions may be gradual or abrupt and may have an irregular character. Therefore, it is difficult to determine the dynamic nature and risk of the escape from the negative lock-in and the pace of the green transformation in Silesia Province.

In light of the above, the aim of the chapter is twofold. First, it is to determine the characteristics of the green transformation of an old industrial region as a response to crisis and decline on the example of Silesia Province. Second, it is to define spatial differences in the eco-innovativeness of industrial enterprises and identify the factors influencing this process in Silesia Province and the associated uncertainties. The research is based on the survey carried out periodically by Statistics Poland related to the implementation of the EIs introduced in production enterprises employing more than 49 people in Silesia Province in the system of NUTS-3 subregions.

The introductory section presents the purpose and focus of the research. The second part includes a review of the literature on the escape of old industrial regions from lock-in scenarios and path development, as well as the issues of the green transformation and EI. The third part briefly presents the socio-economic changes taking place in Silesia Province, with particular emphasis on the processes taking place after 1989. The next section presents the problem of the green transformation in Silesia Province based on the implemented EIs. The chapter ends with a discussion and conclusions regarding the stage of the green transformation in the region.

7.2 Literature Review on the Escape of Old Industrial Regions from the Lock-In

The escape from negative lock-ins that often lead to a deep structural crisis is known to be challenging for old industrial regions (e.g. Grabher 1993). Regions have different capacities to transform and embark on a new development path. As early as 1976, Sydney G. Checkland was one of the first to describe the difficulties in adapting Glasgow's economy after the collapse of the shipbuilding industry there and pointed to the poor adaptability of old industrial regions and the limiting influence of established economic and social structures. As Gwosdz (2014) notes, the transformation of old industrial regions in highly developed countries in the 1980s centered around their deindustrialization. It was most often explained by two concepts: the product life cycle and post-Fordist flexible manufacturing. One of the main achievements of this research was a discovery that old industrial regions have difficulty escaping from lock-ins because of strong unionization, high wages, an unattractive entrepreneurial milieu, a limited number of small companies capable of creating innovations, and pollution (Hall 1985; Saxenian 1994; Storper and Walker 1989).

Later research on changes in old industrial regions was broadened by the issue of over-specialization based on the production of mature technologies (Coenen, Moodysson, and Martin 2015), the concentration of resource-based monostructural clusters of enterprises (Grabher 1993; Hassink and Shin 2005; Schamp 2000). As a result, Hassink (2010) defined old industrial regions as those of the past, in which initial strengths based on geography and networks, such as industrial atmosphere, highly specialized infrastructure, close relations between companies, and strong support by regional institutions, turned into significant barriers to innovation. However, some of the researchers still focused their research on the difficulties dealing with environmental waste and pollution (such as carbon emissions) and underlined that this is one of the main problems of old industrial regions (González-Eguino, Galarraga, and Ansuategi 2012).

As a result, persistent problems in old industrial regions were described by referring to the lock-in concept known from geographical literature (Arthur 1989; Hassink 2010). Lock-in and path dependency concepts have been used broadly and narrowly in evolutionary approaches among economic geographers. The narrow approach is described more extensively by David (2007) and Arthur (1989). In contrast, in the broader approach, the aforementioned authors often refer to the statement "history matters". Hence, Arthur (1989) adopts a process-based view of regional blockage and describes blockage as a gradual process of "stiffening" and increasing inflexibility. When progressive inflexibility occurs, the scope for the escape from lock-in narrows. David (2007) describes the consequences of lock-in in terms of a pool of attraction or equilibrium from which the system can only escape through the destructive impact of an exogenous shock. However, the eventual escape and transition of old industrial regions to new pathways is a problematic process precisely because of the path-dependent nature of technological change (e.g. Smith, Voß, and Grin 2010; Unruh 2000). Therefore, any innovation activities in these regions often follow mature technological trajectories of an incremental character. Efforts to introduce radically new products into the market tend to be limited compared with process optimization and other efficiency-oriented activities (OECD 2019). However, as Gwosdz (2014) notes, the emergence of a crisis in an old industrial region resulting from a long-standing lock-in, combined with local and regional characteristics and the nature of the external institutional environment, creates new opportunities. Such circumstances in combination may lead to a unique conjuncture (critical conjuncture), release old industrial regions from lock-ins, and give rise to a transformation process (David 2007).

The delineation of a new institutional framework resulting from the European Commission climate policy is an opportunity to emerge from the lock-in. In 2019, the European Commission introduced the EGD and programs to complement it, including those in the form of the CEAP of 2015 and its 2020 update. The aim of the first program is a reduction of the pressure on natural resources and the creation of sustainable growth and jobs. The EGD "aims to transform the EU into a fair and prosperous society, with a modern, resource-efficient and competitive economy where there are no net emissions of greenhouse gases in 2050" (European Commission, Directorate-General for Communication 2020, 1). It is also a prerequisite to achieving the EU's 2050 climate neutrality target and halting biodiversity loss. On the other hand, the CEAP announces initiatives along the entire life cycle of products. It is focused on the way products are designed, promotes circular economy processes, encourages sustainable consumption, and aims to ensure that waste is prevented and that the resources used are kept in the EU economy for as long as possible.

The consequences of applying the EGD and CEAP are incentives for implementing EIs in enterprises and the green transformation of old industrial regions. There is vast literature focusing on climate policy and activities to introduce innovative zero-emission technologies, including EIs (e.g. Garrone and Grilli 2010; Hainsch et al. 2021). Most of the authors concentrate on the economic and social role of EIs. The economic role of the EI refers to new products, services, or new business practices necessary to create new business opportunities and benefits for the environment (Arranz et al. 2019; Dangelico 2016; Jové-Llopis and Egarra-Blasco 2018). As for the social consequences of the implementation of EIs, they reduce the environmental impact of consumption or production activities and play a relevant role in the quest for more competitive, environmentally and socially sustainable societies (Carrillo-Hermosilla, Del Río, and Könnölä 2010). Generally, EIs can be defined as those that help open new business niches or generate better competitive behavior, implying a net environmental improvement (Doran and Ryan 2016). In the face of the growing concerns about climate change and owing to the implementation of the European Commission policy to reach the zero-emission target in 2050, it is essential to investigate the EI adoption level in enterprises that helped reduce energy use or CO_2 "footprint", reduce soil, noise, water or air pollution, recycle waste, water, or materials and finally contributed to the replacement of fossil energy with a renewable energy source.

In most of the research focusing on the implementation of innovations, researchers seek drivers standing behind the decision to adopt them. Díaz-García, González-Moreno, and Sáez-Martínez (2015) define two types of factors that influence the introduction of EIs into enterprises: 1) internal (resources and capabilities), and 2) external (public policy and stakeholder impacts). Among the internal factors that influence EI introduction are environmental management (e.g. Rehfeld, Rennings, and Ziegler 2007), equipment renewal (e.g. Cai and Zhou 2014), technological capabilities, internal R&D (Kammerer 2009), cooperation (Mazzanti and Zoboli 2009), skilled personnel (Ghisetti, Marzucchi, and Montresor 2015), or organizational and managerial antecedents (Frondel, Horbach, and Rennings 2008). Despite much research related to the internal factors, Del Río (2009) stresses that a comprehensive perspective on this issue is missing. In particular, the role of resources, competencies, and dynamic capabilities in driving EI has not been researched systematically, even though they are essential drivers of business strategies and innovation performance. In this respect, there is a need to pay attention to raising environmental awareness among employees, skillfully mobilizing participation in environmental training, and implementing EIs (Bianchi 2020). Furthermore, in terms of dynamic capabilities, there is a need to find out how companies adapt an organization's resource base to create a corporate environmental reputation by implementing EIs (Morales-Ray, Martín-Tapia, and Ortiz-de-Mandojana 2019).

With regard to the external factors, Costantini, Crespi, and Palma (2017) and Doran and Ryan (2016) underline that EIs are incentivized not only by the market, which enforces the cost rationalization of the resources used in production processes (energy, water, and materials) but also by environmental regulations and policy (regulatory push/pull effect). The cost rationalization affects the implementation of innovations that help to recycle water, reuse materials, or find new sources of energy (e.g. photovoltaic energy; McCardell 2018). External policies, such as the EGD and CEAP, incentivize the implementation process of this innovation. They imply, for example, tax reduction or financial subsidies for introducing EIs. However, different kinds of EIs contribute differently to a circular economy in the enterprises (Del Río et al. 2016). So far, spatial differences in the influence of internal and external factors on the implementation of EIs in old industrial regions have been poorly studied in the literature (e.g. Antonioli, Borghesi, and Mazzanti 2016).

The impact of these factors on the implementation of EIs may occur with varying intensity in different old industrial regions. This is illustrated by examples from the Ruhr region in Germany (Grabher 1993), North East England (Coenen 2007; Hudson 1994), and Wales (Morgan 2012), where innovations were implemented. Their application in these areas in the 1980s was stimulated by, among other things, regional policies. Birch, Mackinnon, and Cumbers (2010) point to differences between British, German, and French approaches to this policy in old industrial regions. For nearly half a century in the UK, the primary mechanism of structural change in old industrial regions was the policy of attracting outside investors who implemented new technologies and organizational practices. In contrast, in the Federal Republic of Germany and France, change was promoted by using existing regional and national resources and competencies. Diversification into more advanced but related industries to those that played a dominant role in these areas was also promoted (Birch, Mackinnon, and Cumbers 2010; Martin and Sunley 2006).

Martin and Sunley outlined five scenarios for escaping from lock-ins and embarking on a new development path for old industrial regions based on implementing new technologies and innovations (2006, 420; see Table 7.1). As Coenen, Moodysson, and Martin (2015) state, among the escape sources the most likely to occur are "diversification into technologically related industries", which refers to the "transition where an existing industry goes into decline but its core technologies are redeployed and extended to provide the basis of related new industries in the region" (Martin and Sunley 2006, 420). Another relevant option is the "upgrading of existing industries" referring to "the revitalization and enhancement of a region's industrial base through the infusion of new technologies or introduction of new products and services". The authors acknowledge that they are often not mutually exclusive but that different mechanisms may be at work simultaneously and in mutually reinforcing ways. However, whether these escape routes may also counteract each other is left open for debate (Martin and Sunley 2006).

The scenarios presented by Martin and Sunley (2006) dealt with the possible development paths of old industrial regions based on the emergence of new technologies and mainly product or process innovations. These scenarios did

Table 7.1 Escaping a regional lock-in: some possible scenarios.

Source of new path	*Characteristics*
Indigenous creation	New technologies and industries emerge from within the region that have no immediate predecessors or antecedents.
Heterogeneity and diversity	Diversity of local industries, technologies, and organizations promotes constant innovation and economic reconfiguration, avoiding "locking-in" to a fixed structure.
Transplantation from elsewhere	The primary mechanism is the importation of a new industry or technology from elsewhere, which then forms the basis for a new pathway of regional growth.
Diversification into (technologically) related industries	The transition takes place in the event of the decline of an existing industry; however, its core technologies are redeployed and extended to provide the basis for related new industries in the region.
Upgrading of existing industries	A region's industrial base is revitalized and enhanced through the infusion of new technologies or the introduction of new products and services.

Source: based on Lester et al. 2003, 2006 and Martin and Sunley 2006.

not include the processes that might occur when implementing EIs. Therefore, there needs to be more clarity in determining to what extent the scenarios related to escaping a regional lock-in can take a similar course when implementing EIs, or, whether one or more of these scenarios describe the case of an old industrial region of Silesia located in Central Europe.

7.3 Materials and Methods

The research process is based on data on industrial enterprises made available by Statistics Poland, the Center for Science, Technology, and Information Society Statistics. In this regard, the following sources were used:

- national data and data for Silesia Province – obtained primarily from the publicly available reports (covering 2012–2014 and 2018–2020), made available on the website of Statistics Poland, data on the total population of enterprises with 50 or more employees,
- data for the subregions in Silesia Province – obtained from the database prepared exclusively for the authors of the chapter, data on the total population of enterprises with 50 or more employees (because of statistical secrecy, enterprises employing from 10 to 49 people were not taken into account).

The terms used in the above sources, namely industrial enterprises and innovation companies, are defined as follows:

- industrial enterprises are defined and divided according to the Polish Classification of Activities (PKD) 2007, which is consistent with the Statistical Classification of Economic Activities in the European Community (NACE Rev. 2): Section B: Mining and Quarrying, Section C: Manufacturing, Section D: Electricity, Gas, Steam and Air Conditioning Supply, Section E: Water Supply; Sewerage, Waste Management and Remediation Activities,
- innovation activities include all developmental, financial, and commercial activities undertaken by an enterprise intended to result in innovation for an enterprise. Some innovation activities are innovative in their own right; others are not novel but are necessary for implementing innovations.

7.4 Polish Transformation Model of an Old Industrial Region: The Example of Silesia Province

The transformation of old industrial regions in Poland started later than in the countries of the so-called old European Union. This was so because a centrally planned economy had been functioning until 1989 and, as Gwosdz (2014) points out, the socio-economic transformation was initiated after the collapse of the communist system. It resulted in divergent development on a local and regional scale, owing to the following factors: 1) a neoliberal approach to the economy, which favored vital regions at the expense of weak ones; 2) different capabilities of individual metropolitan areas and some areas in western Poland that benefited from connections with Western Europe to attract external capital, and 3) lack of urban policies that could have strengthened the competitiveness of previously invested areas. The process of development divergence also took place in the old industrial regions which include Silesia Province. This voivodeship is one of the largest old industrial regions in Europe (4.5 million people – 12% of Poland's population concentrated in a relatively small population density area – 366 persons per sq km; Statistics Poland 2023), still relying heavily on the extraction and use of fossil fuels as an energy source. It is Poland's most heavily urbanized and second most populous region in the country, which is divided into eight subregions (NUTS 3): Bielsko, Rybnik, Częstochowa, Gliwice, Tychy, Katowice, Sosnowiec and Bytom. The volume of GDP per capita generated in Silesia Province in 2020 was 61.89 thousand PLN (15.87 thousand USD; see Table 7.2), slightly higher than the national average (61.37 thousand PLN; 15.74 thousand USD). The region's specificity is also lower unemployment (3.5% of the unemployment rate, with the national rate being 6.2% at the end of 2020) but also a lower level of education (the proportion of people with tertiary education was 21.4%, compared to 23.1% in Poland).

The European Commission's introduction of the Green Deal in 2019 represents, on the one hand, an opportunity for the transformation and escape from a negative lock-in, but on the other hand, a break with the long-standing tradition of an industrial region as an important raw material and energy base for the whole country. This problem has been widely described in the Polish geographical literature (e.g. Blazyca and Klasik 2003; Gwosdz 2014; Tkocz 2014).

Table 7.2 Socio-economic indices for subregions (NUTS3) in Silesia Province in 2020.

Subregions (NUTS3)	*GDP per capita (in thousand PLN)*	*GDP per capita (in thousand USD)*	*Average gross monthly wages as compared to national average (Poland=100)*	*Share of registered unemployed in working-age population*	*Share of people with higher education**
Bielsko	59.57	15.27	91.3	3.5 %	21.3%
Rybnik	52.00	13.33	106.6	3.3 %	17.1%
Częstochowa	55.02	14.11	84.8	4.4 %	22.4%
Gliwice	74.76	19.17	103.3	3.4 %	21.7%
Tychy	66.71	17.11	90.9	2.6 %	22.0%
Katowice	82.83	21.24	108.8	2.6 %	24.0%
Sosnowiec	55.18	14.15	97.1	4.1 %	22.4%
Bytom	44.94	11.52	90.1	4.3 %	19.3%
Silesia Province	**61.89**	**15.87**	**98.7**	**3.5%**	**21.4%**
Poland	**61.37**	**15.74**	**100**	**4.6%**	**23.1%**

* data from 2021.

Source: Statistics Poland 2020.

The region's association with traditional industries dates back to the 18th century when regular coal mining began in the area. Since then, its extraction increased, and with it came more mines. After the Second World War, the construction of 24 new mines began, including the youngest one, commissioned in 1995. The new mines were primarily intended to ensure a rapid increase in output, so in the post-war years, coal output steadily increased, reaching a peak of 201 million tons in 1979 (Tkocz 2014). Along with the development of the mining industry, its other branches, that is, metallurgy, light industry, and engineering, were developing in the then-functioning Silesia Province. In 1957, there were 3,161 industrial plants in the area, employing 709.9 thousand workers (Grabania 1963, 270).

The political transformation that took place in Poland in 1989 resulted in an increased process of deindustrialization and a profound change in the structure of the economy, especially in the central part of Silesia Province, that is, the Katowice conurbation, Europe's largest polycentric settlement form in terms of the population (Kantor-Pietrag and Krzysztofik 2009; Zagórowska, Rauziński, and Słodczyk 2020). The structural changes were caused not only by the development of new forms of service activities but also by a decline in the number of jobs in the previously dominant industrial activities (Stryjakiewicz 1999). Implemented organizational innovations (e.g. outsourcing) also played an essential role in the transformation. Between 1989 and 2010, non-ferrous metals metallurgy almost wholly disappeared. By the end of 2010, there was only one smelter in the conurbation, and the number of employees had decreased 40-fold, to around 0.5 thousand from over 20 thousand in 1988. The textile industry, traditional in the Zagłębie region and still employing several thousand people in the late 1980s, also disappeared.

Progressive structural changes meant that, at the beginning of the 2010s, the number of people employed in traditional industrial sectors in Silesia Province decreased significantly. The total share of those employed in mining, metallurgy, and energy was at most 12% of the total workforce (in 1988, it was about 40%). Despite this decrease, Silesia Province was still Poland's most extensive coal basin (24 out of 27 domestic hard coal mines were located in the region). The coal industry in the voivodeship employed 74.5 thousand people, which accounted for 89.5% of all employees in this sector in Poland. Coal mining in the region provided 79.7% of domestic hard coal production. While there was a decline in employment in the manufacturing sector as a result of the systemic change, employment in service activities increased. In 2022, the share of the tertiary sector was 62%, 26 percentage points higher than in 1988.

The decline of industry, especially the heavy industry and mining took place in the central and northern part of the core Katowice conurbation, that is, in the centers which formed the industrial core of the region until the 1990s (e.g. the cities of Bytom, Chorzów, and Zabrze). At the same time, the importance of the southern and outer zones of the conurbation increased (e.g. the cities of Katowice and Tychy), where new jobs were created quickly. These areas were characterized by favorable location factors, as they were situated in the A4 motorway corridor, where there were free areas for greenfield investments and better environmental conditions, because of the absence of, *inter alia*, mining damage (Gwosdz 2012, 2014). The division into the former industrial core and the outer zone of the Katowice conurbation is well illustrated by the spatially diversified socio-economic development indicators of Silesia Province by NUTS 3 subregions (see Table 7.2). In this respect, the Katowice subregion with the city of Katowice, which generates the highest value of GDP per capita and the highest wages but also has the highest level of education, is in the best situation.

Despite the good socio-economic situation in the central part of Silesia Province, the Katowice conurbation was diagnosed by the Ministry of Regional Development as one of the two (the other being Eastern Poland) problem areas of national importance because of the coal mining sector and "environmental, social and economic problems of national importance" (Drobniak 2016; Gwosdz 2012). Therefore, the region was included in the process of just transitions of coal-producing regions in the EU, as a part of the European Commission European Green Deal policy. Since 2022 the Regional Plan for Just Transition (based primarily on European funds) has been implemented, with a budget of approximately 6.5 bn PLN (1.6 bn USD current prices) until 2030 (cf. European Commission, Directorate-General for Communication 2020). The implementation of this program offers a real opportunity for Silesia Province to escape from the lock-in and start the green transformation. The guidelines of the European Commission's 2020 CEAP, which set out activities related to innovative solutions for a circular economy, also play an essential role.

7.5 Results: The Escape of Silesia Province from the Lock-In through Green Transformation

The degree of the green transformation in Silesia Province depends on the region's ability to adopt new solutions and innovations to reduce pollution and recycle waste, among other things. However, without determining the level of innovativeness of industrial enterprises in general, it is difficult to study their ability to implement EIs. Hence, this part of the chapter adopts this approach and then presents the eco-innovativeness level of Silesia Province enterprises by NUTS-3 subregions. The final part presents internal and external factors that influenced the innovative activity of the studied enterprises.

The research on the level of innovative activity in industrial enterprises with more than 49 employees in Silesia Province in 2014 and 2020 indicated rising figures (see Table 7.3). The share of enterprises implementing innovations increased from almost 22% to 36.6%. However, during this period, there was a process of "catching up" of Silesia Province's innovativeness by other regions in Poland. While in 2014, the analyzed region showed a higher level of innovativeness than the whole of Poland, by 2020 this difference became almost imperceptible. This change, however, was associated with a significant increase in the innovative activity of manufacturing companies in Poland as a whole in

Table 7.3 Innovation and EI activity of industrial enterprises in Silesia Province in 2018–2020 in percent.

Type of innovation activity	*Poland*			*Silesia Province*		
	Year		*Change (2014–2020)*	*Year*		*Change (2014–2020)*
	2014	*2020*		*2014*	*2020*	
Innovation of active industrial enterprises in years 2014–2020	18.6	36.7	18.1	21.9	36.6	14.7
Eco-innovation active industrial enterprises in years 2014–2020	21.1	35.0	13.9	35.8	37.3	1.5

Source: Statistics Poland 2015, 2021.

the period under consideration, which was mainly due to the increase in support for the innovative activity of companies by EU funds and the policies pursued by regional and local authorities (Klemens and Heffner 2017). As a result, highly processed and innovative goods (e.g. electronic equipment, buses, 3D printers, and biotechnology) have started to play an increasingly important role in the structure of Polish exports (Ambroziak et al. 2020). At the same time, the weakening position of the voivodeship among innovation leaders may be evidenced by its low position in terms of innovation efficiency, measured by the value of expenditure on research and development activities within the innovative activity of enterprises. In terms of the amount of funds earmarked for this purpose in 2020, Silesia Province was in the lead nationwide (2,794.3 million PLN/1,012.4 million USD) after the provinces in which the largest Polish cities are located: Warsaw: Mazovia Province, and Krakow: Lesser Poland (Małopolskie) Province, while its position in terms of the level of innovation was at the national average. Moreover, as indicated by Statistics Poland report (2021), Silesian enterprises in comparison with other regions are more willing to cooperate within the framework of innovation activities (3rd place countrywide). This cooperation meant active participation in joint projects with other enterprises or non-commercial institutions (Oslo Manual 2018). The high level of cooperation between enterprises in the implementation of innovative projects results, among other things, from the implementation of two regional innovation strategies for Silesia Province, the first one for 2003–2013 and the second one for 2013–2020 (Regional Innovation Strategy of the Śląskie Voivodeship for 2003–2013; Regional Innovation Strategy of the Śląskie Voivodeship for 2014–2020). In both of these documents, building cooperation networks, clusters, and scientific-industrial consortia was adopted as one of the main objectives in developing new technologies and increasing the innovativeness of local entities.

The activity of large industrial enterprises in the adoption of EI plays an important role in the green transformation. In this respect, the activity of the surveyed entities in Silesia Province was high, as both in 2014 and 2020 exceeded the average value for the whole country (21.1% and 35.0%, respectively; see Table 7.3). However, the dynamic of this process in the considered period of six years was not high, which may indicate the occurrence of a barrier that is difficult to overcome in the long term. Such a barrier in the case of Silesia Province may be informal restrictions, related to the routines, conventions, and traditions of the region's coal mining heritage that, as is well-known, change slowly over time (Martin and Sunley 2006).

When considering the total innovative activity of enterprises by NACE sections, enterprises that are part of the Manufacturing Section were the most efficient (59.5% innovative active enterprises, 18.9% increase of activity). This is due to the implementation of innovations related to advanced production systems contributing to more flexible processes, including reducing fixed costs and shortening the delivery time of new products to the market (see Table 7.4). In addition, manufacturing companies from Silesia Province have in their export portfolio the increasing number of new and innovative parts and accessories for the production of passenger vehicles, of which about 20% are sold on the EU market (Regional Innovation Strategy of the Śląskie Voivodeship 2030).

On the other hand, the entities that were most active in terms of implementing EIs were those that relied on natural energy resources in their activities and which are part of the Electricity, Gas, Steam, and Air Conditioning Supply Section as well as the Water Supply; Sewerage, Waste Management, and

Table 7.4 Innovation-active and eco-innovation-active industrial enterprises in Silesia Province by NACE divisions in 2014 and 2020.

NACE sections	*Innovation-active industrial enterprises*			*Eco-innovation-active industrial enterprises*		
	Year		*Change (2014–2020)*	*Year*		*Change (2014–2020)*
	2014	*2020*		*2014*	*2020*	
Mining and Quarrying	30.6	42.9	12.3	11.7	26.4	14.7
Manufacturing	40.6	59.5	18.9	18.6	28.5	9.9
Electricity, Gas, Steam, and Air Conditioning Supply	47.2	51.4	4.2	27.9	47.2	19.3
Water Supply; Sewerage, Waste Management, and Remediation Activities	38.0	40.4	2.4	37.1	42.8	5.7

Source: Statistics Poland 2023.

Remediation Activities Section. The enterprises from the former section implemented innovative solutions to obtain renewable energy from photovoltaic or wind farms or small hydropower plants. One example of such activity is opening one of Poland's largest photovoltaic power plants by one of the country's leading energy suppliers, Tauron, in Mysłowice (subregion NUTS 3 Katowice) in 2023. Its capacity will eventually reach 100 megawatts (Slaskibiznes 2023). There has also been an increase in the use of energy from innovative solutions introduced in biogas plants, for example, in the municipality of the Koszęcin–Bytom subregion, one of the first container biogas plants in Poland was built, capable of producing about 20 kilowatts of electricity per hour (Malopolska 2023). Moreover, enterprises from the latter section, Water Supply; Sewerage, Waste Management, and Remediation Activities, have innovated to reduce waste production and reuse, as well as to close the water cycle in production systems, analyze material flow, and assess the product life cycle. The action taken by the Katowice city authorities to close the water cycle may serve as an example in this respect. In this way, the bodies responsible for water and wastewater management are tasked with introducing innovative solutions for reducing water consumption and closing the water cycle (Portalkomunalny 2023). Some of the measures taken have already been initiated as part of the programs introduced by the European Green Deal, under the CEAP of 2015, the Green Paper "On Energy Security" of 2000, among others, and from programs and legal documents operating at local and regional levels, for example, the Decree of the Minister of Economy of 2000 on the obligation to purchase electricity from unconventional and renewable sources.

The system transformation after 1989 was to a large extent associated with catching up with Western European economies (Stryjakiewicz, Burdack, and Egedy 2010) and closing the existing gaps in the implementation of investments related to the construction or renovation of, inter alia, the basic sewage and water supply infrastructure. Although the period to which the analysis relates is the quarter of a century away from the collapse of the communist system, the type of EIs introduced, described especially at the beginning of the study, indicates a continuous process of filling in the existing infrastructure gaps. This involves the introduction of EIs related to recycled waste, water, or materials for own use or sale. Innovations in this area accounted for the dominant share of technological solutions introduced in 2014 (see Figure 7.1). The improvement to the air quality and reduction of CO_2 emissions in Silesia Province was another issue related to the long-lasting legacy of traditional energy industries based on the extraction and combustion of fossil fuels. Hence, in 2020, the most significant of the EIs introduced in the region's enterprises was the improvement in energy consumption and reduction of the CO_2 "footprint" (i.e. the reduction of total CO_2 emissions). This type of innovation was introduced by 1/3 of all innovation-active enterprises. Although the implementation rate of EIs replacing the share of fossil energy with renewable energy sources was low during the analyzed period, it increased significantly from 2014 to 2020. The noticeable increase in the role of these EIs illustrates the

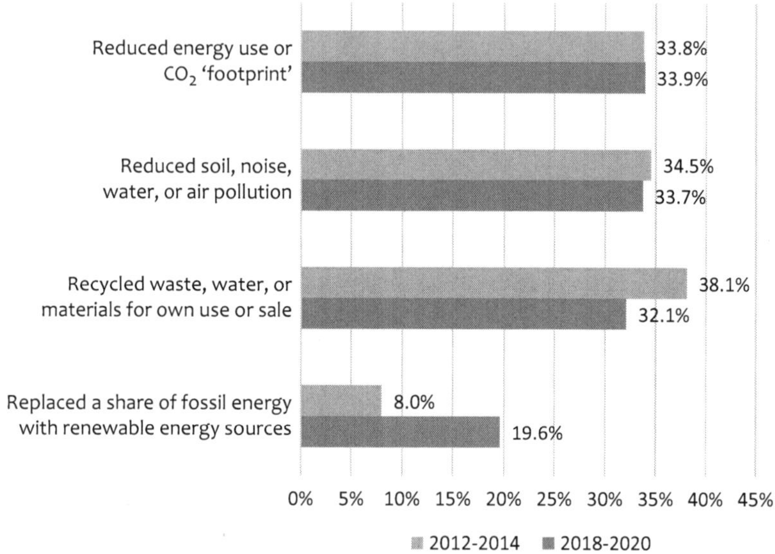

Figure 7.1 Industrial enterprises in Silesia Province which introduced EIs bringing environmental benefits obtained during the period of manufacturing goods or services by the enterprise (as the share of innovation-active industrial enterprises).

Source: Statistics Poland 2023.

future of resource-based green energy, mainly solar and wind. The rising cost of fossil fuel-based energy will stimulate this growth.

A spatial analysis of the variation in the activity of enterprises in implementing EIs was carried out in the form of a cartographical data presentation, that is, a cartogram. For this purpose, the average value of this activity was determined, 35.8% for 2014 and 37.3% for 2020. The boundaries of the range into which NUTS 3 subregions with an average level of activity in the EI implementation were included were determined by adding or subtracting the value of the standard deviation, which was 9.5 for 2014 and 6.2 for 2020. On the other hand, the remaining ranges defined as low or high denote the values obtained by the subregions above or below the range defined as medium. The result of the analysis is shown in Figure 7.2.

In 2014, most of the regions fell within the range defining average activity in terms of EIs implemented. In the Częstochowa subregion, innovation activity was the lowest because of the small number of large industrial enterprises and the loss of its original industrial character. On the other hand, the Rybnik subregion, located in the western part of the voivodeship, was distinguished in 2014 by an above-average level of business activity in the implementation of EIs, in particular in the recycling of waste, water, or materials for own use or sale. This section was quite strongly represented in the subregion by large and some of the most profitable waste collection and recycling enterprises in the

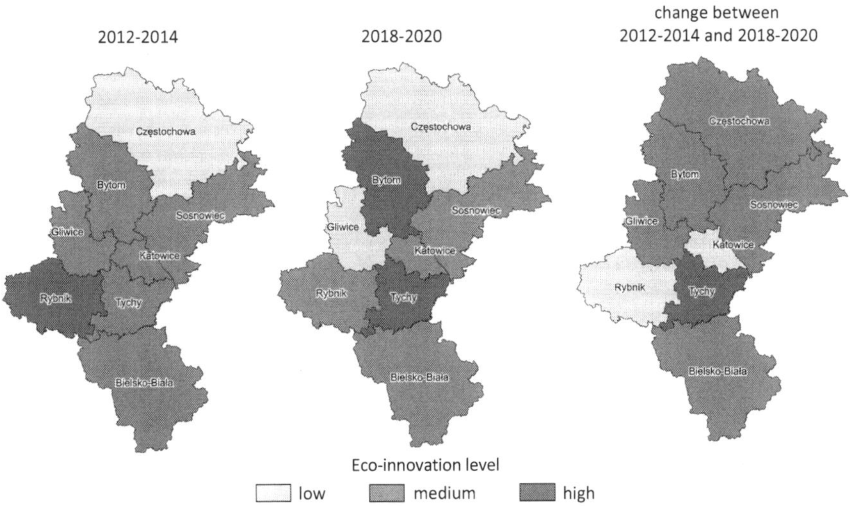

Figure 7.2 EI level of manufacturing enterprises (from Sections B to E of NACE) with more than 49 employees in Silesia Province by NUTS3 subregions in 2014 and 2020.

Source: Statistics Poland 2023.

region, which had the capital resources to implement innovative solutions (Rybnik 2023).

On the other hand, in 2020, the highest activity in terms of EI implementation was recorded in Bytom and Tychy subregions. In the first of the analyzed subregions, the innovative activity took place in each of the sections under consideration, with a dominant role of industrial processing, which is concentrated, inter alia, in the recultivated areas of former mining and metallurgical plants where the Bytom Industrial Park (Bytomski Park Przemysłowy) has been operating since 2008. In this subregion, companies involved in the reclamation of areas left after coal mines that no longer exist in the area are highly active in implementing EIs. More than high activity was shown by enterprises from Section D related to the production and supply of electricity, gas, steam, hot water, and air for air-conditioning systems. Among these enterprises, the Miechowice Combined Heat and Power Plant, operating in the area, was very active, introducing EIs to reduce energy intensity or CO_2 emissions. In addition, high activity in implementing EI was observed in the Tychy subregion, which also saw the highest increase in the percentage of enterprises that implemented EIs between 2014 and 2020 (12.7%). Industrial enterprises located in this subregion implemented each of the considered types of EIs for environmental protection. These entities included Tyskie Browary Książęce (Prince's Tychy Brewery), owned by one of Poland's largest brewing companies, which is part of the Japanese brewing conglomerate Asahi. The company has primarily implemented solutions based on closing the water circuit as part of its production processes.

Tychy is also home to the highly innovative automotive plants of the Fiat Corporation and General Motors, introducing EIs in the field of waste recycling.

In the analysis related to the implementation of innovations, the factors influencing the decision to apply the innovation play a vital role (Męczyński 2007). The survey revealed that enterprises in Silesia Province attached greater importance to external factors in implementing EIs (see Figure 7.3). Surprisingly, among these factors, in 2014 and 2020, government grants, subsidies, or other financial incentives for introducing innovations with environmental benefits were not the most important for entrepreneurs implementing EIs. A more significant role was played by the costs of energy, water or material costs, and existing environmental regulations; environmental taxes, charges, or fees are considered particularly important in this regard. On the other hand, companies also take into account internal factors in implementing EI, such as: improving an enterprise's reputation or voluntary actions, or initiatives for environmental good practice within the enterprise's sector. The importance of most of the factors motivating the implementation of EIs in enterprises has declined during the period under review. A particularly large decline is observed among internal factors, by an average of five percentage points. What implies the shift from the company's driven innovative activity to the external incentives, which is confirmed by the survey, is the increasing role of government grants, subsidies, or other financial incentives for introducing innovations with environmental benefits (from 21.4% in 2014 to 23.9% in 2020).

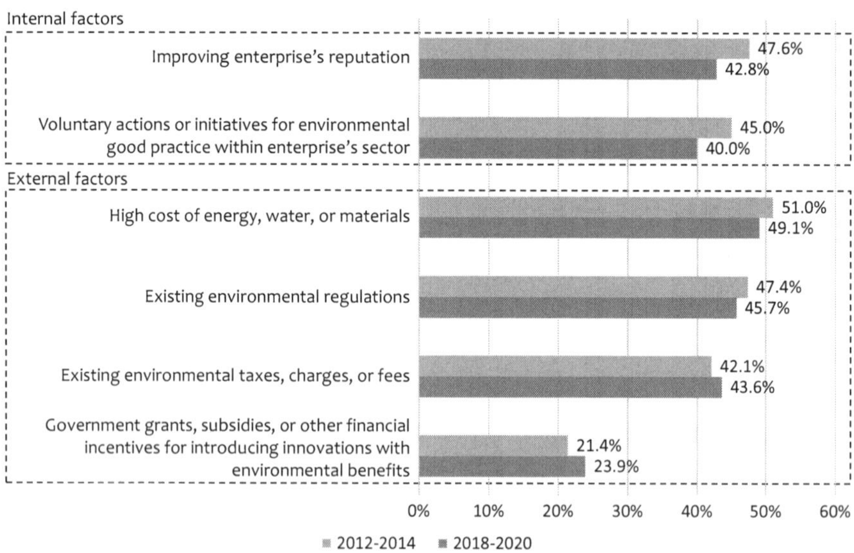

Figure 7.3 Industrial enterprises in Silesia Province which rated the importance of internal and external factors in driving an enterprise's decisions to introduce EIs as "important" (the share of innovation-active industrial enterprises).

Source: Statistics Poland 2023.

The European Green Deal and the approved 2022 Regional Plan for Just Transition (based primarily on European funds) are the answer to the growing expectation of entrepreneurs with regard to external incentives in the implementation of innovations. It identifies social, economic, and environmental challenges and needs for economic branching, worker retraining, and environmental rehabilitation, including through EIs. Under the plan, the remaining coal mines in Silesia Province will be gradually closed, resulting in a decline of 7.7 million tons in coal production, from 30 million tons per year in 2021 to 23 million tons annually in 2030. As a result, by 2030, there will be a decline of nearly 80% in electricity production from conventional coal. The planned reduction in coal-fired power generation will result in a 62.1% reduction in CO_2 (from 13.2 million tons in 2019 to 5 million tons in 2030).

The primary source of funding for the goals of the Just Transition Territorial Plan is EU funds for 2021–2027, that is, the Just Transition Fund supplemented by grants from the European Regional Development Fund and the European Social Fund. From these sources, it is primarily planned to finance incentives, including grants, loans, awards, and guarantees for investments for funding aimed at the holistic green transformation of the region. However, there needs to be more dedicated legislation for Silesia Province to accelerate the change.

7.6 Discussion and Conclusions

The results of the analysis presented in the chapter made it possible to answer the two objectives set out in the introduction. As regards the first one, which concerns defining the characteristics of the green transformation in one of the European old industrial regions, that is, Silesia Province, the concept of lock-in and path dependency was adopted in broad terms. From that perspective, the understanding of the concept is reduced to the description that "history matters" and that, generally, the phenomena and processes occurring in the past condition the present processes (Arthur 1989; David 2007). By framing the process of the green transformation in Silesia Province in this way, it has been possible to see the potential of the technological changes of the region based on adopting EIs to escape the negative lock-in. The problem of the negative lock-in of Silesia Province is associated with the still high operation of industries that use fossil fuels and contribute to environmental degradation. Despite the initial phase of this transformation, it is possible to identify the characteristics of the province's escape from the lock-in and the entry into the new development path. Based on the survey related to the EI activity and the authors' desk research of the statistical data and literature related to the dynamic structure of the voivodeship regional economy, the presented analysis results provide evidence that the escape is possible. The escape process is characterized by the two interrelated and complementary scenarios described in the literature by Martin and Sunley (2006). The first is the diversification into (technologically) related industries and the second is upgrading existing industries. The first scenario of escape and get on the new path of development for Silesia Province is based on the solid evidence that most of the old mining enterprises went into decline (the total share of employees in mining, metallurgy, and energy did not exceed 12% of the total workforce, while in 1988 it was around 40%). However, fossil fuel-based energy companies that still exist (e.g. Koksownia Radlin in the Rybnik subregion) implement various new technologies and EIs to protect the environment and reduce the emission of CO_2. Those revitalized enterprises have also started to cooperate with the related new companies, including start-ups (e.g. enterprises involved in the reclamation of the degraded soil and land that formally belonged to the mining enterprises). Moreover, the existing power distribution enterprises based on fossil fuels cooperate with the newly established solar energy producers and provide the energy based on a mix of sources (traditional non-renewable sources and modern renewable sources).

The second scenario is based on the upgrading of the existing industries. This process occurs through the revitalization and enhancement of the region's industrial base through the implementation of EIs and mainly within the two sections of NACE classification: 1) Electricity, Gas, Steam, and Air Conditioning Supply, and 2) Water supply; Sewerage, Waste Management, and Remediation Activities. Various EIs were introduced in both of these sections. However, as the socio-economic changes that started in Silesia Province were introduced late compared to the relevant Western European old industrial regions, the EIs implemented were related to basic infrastructure, for example, innovations that improve the recycling of waste, water, or materials for own use or sale.

The pace of implementing EIs also characterized the green transformation in Silesia Province. Although the analysis covers a relatively short period from 2014 to 2020, which is mainly before the European Green Deal and when climate policy was built on less ambitious aims, it provides some records on the process. They show that more than 1/3 of the studied industrial enterprises were actively implementing EIs in both investigated moments, and the pace of the change was slow (a 1.5% increase occurred). This slow pace process describes the incremental, no disruptive nature of the change. Therefore, given the pace of this process, Silesia Province could be regarded as a typical old industrial region, like many in Western Europe, which struggles to escape the lock-in and get on the new path of development (Coenen, Moodysson, and Martin 2015; Hassink and Shin 2005; Stryjakiewicz, Burdack, and Egedy 2010).

The second aim of the chapter was to determine spatial differences in the eco-innovativeness of industrial enterprises and to identify factors that influence this process in Silesia Province. The eco-innovativeness in the region is affected to a large extent by the pace of the socio-economic changes that occurred after the collapse of communism in Poland in 1989, as well as Poland's accession to the European Union in 2004. Especially the latter opened up the possibility for obtaining the European Funds to restructure the traditional industry in Silesia Province. The spatial differences in the eco-innovativeness mostly correspond to the general innovativeness of the region as well as the socio-economic development. However, the Katowice subregion is located in the central part of the province and, as one of the most populous old industrial regions in Europe, represents an average level of eco-innovativeness, despite the concentration of academic institutions and a well-developed climate of cooperation. However, because of the socio-economic processes occurring since 1989, the role of the Katowice subregion as an industrial center declined in favor of services. Therefore, the subregion concentrates on other types of innovations. The most eco-innovative subregions are those where most of the traditional and fossil fuel-based industries are located (the subregions of Bytom, Rybnik, and Tychy), which benefit from many forms of the EU funds, including the recently available regional Just Transition Fund. Therefore, the green transformation is positively affecting the core industries responsible for the emission of CO_2. However, it is too early to determine whether the escape of Silesia Province from the lock-in was successful.

References

Ambroziak, Ł., J. Chojna, J. Gniadek, A. Krawczyk, K. Marczewski, and J. Sawulski, J. 2020.

Antonioli, D., S. Borghesi, and M. Mazzanti. 2016. Are regional systems greening the economy? Local spillovers, green innovations and firms' economic performances. *Economics of Innovation and New Technology* 25 (7):692–713. doi: 10.1080/10438599.2015.1127557

Arranz, N., M. F. Arroyabe, A. Molina-García, and J. F. de Arroyabe. 2019. Incentives and inhibiting factors of eco-innovation in the Spanish firms. *Journal of Cleaner Production* 220:167–76. doi: 10.1016/j.jclepro.2019.02.126

Arthur, W. B. 1989. Competing Technologies. Increasing Returns, and Lock-In by Historical Events. *The Economic Journal* 99 (394):116–31. doi: 10.2307/2234208

Bianchi, G. 2020. *Sustainability competences*. Luxembourg: Publications Office of the European Union, JRC123624. doi: 10.2760/200956

Birch, K., D. MacKinnon, and A. Cumbers. 2010. Old industrial regions in Europe: a comparative assessment of economic performance. *Regional Studies* 44 (1):35–53. doi: 10.1080/00343400802195147

Blazyca, G., and A. Klasik. 2003. Transforming industrial dynamo into strong regional economies – a comparison of the West of Scotland and Upper Silesia. In *Restructuring regional and local economies. Towards a comparative study of Scotland and Upper Silesia*, ed. G. Blazyca. London: Ashgate. doi: 10.4324/9781315194448

Boschma, R., and J. Lambooy. 1999. Evolutionary economics and economic geography. *Journal of Evolutionary Economics* 9:411–29. doi: 10.1007/s001910050089.

Cai, W. G., and X. L. Zhou. 2014. On the drivers of eco-innovation: empirical evidence from China. *Journal of Cleaner Production* 79: 239–48.

Carrillo-Hermosilla, J., P. Del Río, and T. Könnölä. 2010. Diversity of Eco-Innovations: Reflections from Selected Case Studies. *Journal of Cleaner Production* 18 (10/11): 1073–83. doi: 10.1016/j.jclepro.2010.02.014

Checkland, S. G. 1976. *Upas Tree-Glasgow, 1875–1975*. Rowman & Littlefield Publishers, Incorporated.

Coenen, L. 2007. The role of universities in the regional innovation systems of the north east of England and Scania, Sweden: providing missing links? *Environment and Planning C: Government and Policy* 25:803–21. doi: 10.1068/c0579

Coenen, L., J. Moodysson, and H. Martin. 2015. Path Renewal in Old Industrial Regions: Possibilities and Limitations for Regional Innovation Policy. *Regional Studies* 49 (5):850–65. doi: 10.1080/00343404.2014.979321

Costantini, V., F. Crespi, and A. Palma. 2017. Characterizing the policy mix and its impact on eco-innovation: A patent analysis of energy-efficient technologies. *Research Policy* 46 (4):799–819. doi: 10.1016/j.respol.2017.02.004

Dangelico, R. M. 2016. Green Product Innovation: Where we are and Where we are Going. *Business Strategy and the Environment* 25: 560–76. doi: 10.1002/bse.1886

David, P. A. 2007. Path dependence: a foundational concept for historical social science. *Cliometrica* 1 (2):91–114. doi: 10.1007/s11698-006-0005-x

Del Río, P. 2009. The empirical analysis of the determinants for environmental technological change: A research agenda. *Ecological Economics* 68 (3):861–78. doi: 10.1016/j.ecolecon.2008.07.004

Del Río, P., J. Carrillo-Hermosilla, T. Könnölä, and M. Bleda. 2016. Resources, capabilities and competences for eco-innovation. *Technological and Economic Development of Economy* 22 (2):274–92. doi: 10.3846/20294913.2015.1070301

Díaz-García, C., Á. González-Moreno, and F. J. Sáez-Martínez. 2015. Eco-innovation: insights from a literature review. *Innovation* 17 (1):6–23. doi: 10.1080/14479338.2015.1011060.

Doran, J., and G. Ryan. 2016. The effectiveness of R&D and external interaction for innovation: Insights from quantile regression. *Economic Issues Journal Articles, Economic Issues* 21 (1):47–65.

Drobniak, A. 2016. Ekonomiczne koncepcje rozwoju w kontekście rewitalizacji miast - studia przypadków [Economic concepts of development in the context of urban regeneration - case studies]. *Prace Naukowe Uniwersytetu Ekonomicznego we Wrocławiu* 6 (432):27–41. doi: 10.15611/pn.2016.432.03

European Commission, Directorate-General for Communication. 2020. *Circular economy action plan: for a cleaner and more competitive Europe*. Brussels: Publications Office of the European Union.

Friedmann, J. 1992. *Empowerment. The politics of an alternative development*. Oxford: Basil Blackwell.

Frondel, M., J. Horbach, and K. Rennings. 2008. What triggers environmental management and innovation? Empirical evidence for Germany. *Ecological Economics* 66 (1):153–60. doi: 10.1016/j.ecolecon.2007.08.016

Garrone, P., and L. Grilli. 2010. Is there a relationship between public expenditures in energy R&D and carbon emissions per GDP? An empirical investigation. *Energy Policy* 38 (10):5600–13. doi: 10.1016/j.enpol.2010.04.057

Ghisetti, C., A. Marzucchi, and S. Montresor. 2015. The open eco-innovation mode. An empirical investigation of eleven European countries. *Research Policy* 44 (5):1080–93. doi: 10.1016/j.respol.2014.12.001

González-Eguino, M., I. Galarraga, and A. Ansuategi. 2012. The future of old industrial regions in a carbon-constrained world. *Climate Policy* 12 (2):164–86. doi: 10.1080/14693062.2011.605707

Grabania, M. 1963. *Regiony przemysłowe województwa katowickiego* [Industrial regions of the Katowice Province]. Katowice: Wydawnictwo Śląsk.

Grabher, G. 1993. *The embedded firm. On the socioeconomics of industrial network*. London: Routledge: 255–77.

Gwosdz, K. 2012. Baza ekonomiczna i specjalizacja funkcjonalna miast konurbacji katowickiej po dwu dekadach transformacji [Economic base and functional specialisation of the cities of the Katowice conurbation after two decades of transformation]. *Acta Geographica Silesiana* 11:15–29.

Gwosdz, K. 2014. *Pomiędzy starą a nową ścieżką rozwojową: mechanizmy ewolucji struktury gospodarczej i przestrzennej regionu tradycyjnego przemysłu na przykładzie konurbacji katowickiej po 1989 roku* [Between old and new development paths: mechanisms of economic and spatial structure evolution in a region of traditional industry on the example of the Katowice conurbation after 1989]. Kraków: Instytut Geografii i Gospodarki Przestrzennej Uniwersytetu Jagiellońskiego.

Hainsch, K., T. Burandt, K. Löffler, C. Kemfert, P. Y. Oei, and C. von Hirschhausen. 2021. Emission pathways towards a low-carbon energy system for Europe: a model-based analysis of decarbonization scenarios. *The Energy Journal* 42 (5):41–66. doi: 10.5547/01956574.42.5.khai

Hall, S. 1985. Authoritarian populism: A reply to Jessop et al. *New left review* 151 (1):115–23.

Hassink, R. 1993. Regional innovation policies compared. *Urban Studies* 30 (6):1009–24. doi: 10.1080/00420989320080921

Hassink, R. 2010. Locked in decline? On the role of regional lock-ins in old industrial areas. In *Handbook of evolutionary economic geography*, ed. R. Boschma, and R. Martin, 450–68. London: Edward Elgar.

Hassink, R., and D. H. Shin. 2005. South Korea's shipbuilding industry: From a couple of Cathedrals in the desert to an innovative cluster. *Asian Journal of Technology Innovation* 13 (2):133–55. doi: 10.1080/19761597.2005.9668611

Henderson, S. R. 2015. Transforming old industrial regions: Constructing collaboration within the Black Country, England. *Geoforum* 60:95–106. doi: 10.1016/j.geoforum.2015.01.010

Hudson, R. 1994. New production geographies? Reflections on changes in the automobile industry. *Transactions of the Institute of British Geographers* 19:331–45. doi: 10.2307/622326

Jakobsen, S. E., E. Uyarra, R. Njøs, and A. Fløysand. 2022. Policy action for green restructuring in specialized industrial regions. *European Urban and Regional Studies* 29 (3):312–31. doi: 10.1177/09697764211049116

Jové-Llopis, E., and A. Segarra-Blasco. 2018. Eco-innovation strategies: A panel data analysis of Spanish manufacturing firms. *Business Strategy and the Environment* 27: doi:10.1002/bse.2063

Kammerer, D. 2009. The effects of customer benefit and regulation on environmental product innovation: Empirical evidence from appliance manufacturers in Germany. *Ecological Economics* 68 (8–9):2285–95. doi: 10.1016/j.ecolecon.2009.02.016

Kantor-Pietraga, I., and R. Krzysztofik. 2009. Rozwój demograficzny miast na obszarze województwa ślaskiego od XIII do XXI wieku [Demographic development of towns in the Silesia Province from the 13th to the 21st century]. *Prace Wydzialu Nauk o Ziemi Uniwersytetu Slaskiego*. Sosnowiec: Uniwersytet Śląski.

Klemens, B., and K. Heffner. 2017. Czynniki innowacyjności polskich regionów [Innovation factors of Polish regions]. *Research Papers of Wrocław University of Economic* 467:187–201. doi: 10.15611/pn.2017.467.16

Lester, A. 2006. Colonial Networks, Australian Humanitarianism and the History Wars. *Geographical Research* 44:229–41. doi: 10.1111/j.1745-5871.2006.00387.x

Lester, D., J. Parnell, and S. M. Carraher 2003. Organizational life cycle: A five-stage empirical scale. *International Journal of Organizational Analysis* 11:339–54. doi: 10.1108/eb028979

Machiba, T. 2010. Eco-innovation for enabling resource efficiency and green growth: development of an analytical framework and preliminary analysis of industry and policy practices. *International Economics and Economic Policy* 7:357–70. doi: 10.1007/s10368-010-0171-y

Malopolska [Lesser Poland]. 2023. Accessed December 03, 2023. https://www.malopolska.pl/

Martin, R., and P. Sunley. 2006. Path dependence and regional economic evolution. *Journal of Economic Geography* 6 (4):395–437. doi: 10.1093/jeg/lbl012

Mazzanti, M., and R. Zoboli. 2009. Embedding environmental innovation in local production systems: SME strategies, networking and industrial relations: evidence on innovation drivers in industrial districts. *International Review of Applied Economics* 23 (2):169–95. doi: 10.1080/02692170802700500

McCardell, S. 2018. *Energy Effectiveness: Strategic Objectives, Energy and Water at the Heart of Enterprise*. Springer.

Męczyński, M. 2007. *Przestrzenne zróżnicowanie i dyfuzja technologii informacyjno-komunikacyjnych [Spatial differentiation and diffusion of information and communication technologies]*. Poznań: Bogucki Wydawnictwo Naukowe.

Morales-Raya, M., I. Martín-Tapia, and N. Ortiz-de-Mandojana. 2019. To be or to seem: The role of environmental practices in corporate environmental reputation. *Organization and Environment* 32 (3):309–30. doi: 10.1177/1086026617753154

Morgan, M. S. 2012. *The world in the model: How economists work and think*. Cambridge University Press. doi: 10.1017/CBO9781139026185

OECD 2019. *Regions in Industrial Transition: Policies for People and Places*. Paris: OECD Publishing.

Oslo Manual. 2018. *Guidelines for Collecting, Reporting and Using Data on Innovation. The Measurement of Scientific, Technological and Innovation Activities*. 4th ed. OECD/Eurostat.

Portalkomunalny [Municipal portal]. 2023. Accessed December 03, 2023. https://portalkomunalny.pl/

Regional Innovation Strategy of Silesia Voivodeship 2030. 2020. Katowice: Urząd Marszałkowski Województwa Śląskiego.

Regional Innovation Strategy of the Silesia Voivodeship for 2003–2013. 2002. Katowice: Urząd Marszałkowski Województwa Śląskiego.

Regional Innovation Strategy of the Silesia Voivodeship for 2014–2020. 2003. Katowice: Urząd Marszałkowski Województwa Śląskiego.

Rehfeld, K. M., K. Rennings, and A. Ziegler. 2007. Integrated product policy and environmental product innovations: An empirical analysis. *Ecological Economics* 61 (1):91–100. doi: 10.1016/j.ecolecon.2006.02.003

Rybnik. 2023. Accessed November 29, 2023. www.rybnik.naszemiasto.pl

Saxenian, A. L. 1994. *Regional Advantage: Culture and Competition in Silicon Valley and Route 128*, Cambridge: Harvard University Press.

Schamp, E. W. 2000. *Decline and renewal in industrial districts: exit strategies of SMEs in consumer goods industrial districts of Germany. The networked Firm in a Global World. Small firms in new environments*. Aldershot: Ashgate: 257–82.

Slaskibiznes [Silesian business]. 2023. Accessed December 03, 2023. https://www.slaskibiznes.pl/

Smith A., J. P. Voß, and J. Grin. 2010. Innovation studies and sustainability transitions: The allure of the multi-level perspective and its challenges. *Research Policy* 39:435–48. doi: 10.1016/j.respol.2010.01.023

Statistics Poland. 2015. Działalność innowacyjna przedsiębiorstw w latach 2012-2014. Urząd Statystyczny w Szczecinie [Innovative activities of enterprises in 2012-2014]. Statistical Office in Szczecin. GUS. Warszawa.

Statistics Poland. 2020. Accessed November 20, 2023. https://bdl.stat.gov.pl

Statistics Poland. 2021. Działalność innowacyjna przedsiębiorstw w latach 2018–2020 [Innovation activities of companies 2018-2020]. Statistical Office in Szczecin. GUS. Warszawa, Szczecin.

Statistics Poland. 2023. Data prepared by the Statistics Poland in 2023, at the request of the authors of the paper.

Storper, M., and R. Walker. 1989. *The capitalist imperative: territory, technology and industrial growth*. Oxford: Basil Blackwell.

Stryjakiewicz, T. 1999. *Adaptacja przestrzenna przemysłu w Polsce w warunkach transformacji* [Spatial adaptation of industry in Poland under conditions of transition]. Poznań: Wydawnictwo Naukowe Uniwersytet im. Adama Mickiewicza w Poznaniu.

Stryjakiewicz, T., J. Burdack, and T. Egedy. 2010. Institutional Change And New Development Paths: Budapest, Leipzig, Poznan, Riga And Sofia. In *Making Competitive Cities*, ed. S. Musterd, and A. Murie, 93–112. Oxford: Wiley-Blackwell.

Tkocz, M. 2014. Zmiany gospodarcze Zagłębia Dąbrowskiego w okresie funkcjonowania Polski w Unii Europejskiej [Economic changes in the Zagłębie Dąbrowskie region during Poland's membership in the European Union], In *10 lat Zagłębia Dąbrowskiego w Unii Europejskiej* [10 years of the Zagłębie Dąbrowskie region in the European Union], ed. M. Barański, 42–46. Sosnowiec: Wydawnictwo Zagłębiowska Oficyna Wydawnicza.

Tödtling, F. 1992. Technological change at the regional level: the role of location, firm structure, and strategy. *Environment and Planning A* 24 (11):1565–84. doi: 10.1068/a241565

Trippl, M., and A. Otto. 2009. How to turn the fate of old industrial areas: a comparison of cluster-based renewal processes in Styria and the Saarland. *Environment and Planning A: Economy and Space* 41 (5):1217–33. doi: 10.1068/a4129

Unruh, G. 2000. Understanding Carbon Lock-In. *Energy Policy* 28:817–30. doi: 10.1016/S0301-4215(00)00070-7

Wiedermann, K. 2007. Regionalne efekty mnożnikowe rozwoju przemysłu motoryzacyjnego w aktywizacji gospodarczej województwa śląskiego [Regional multiplier effects of automotive industry development in the economic activation of the Silesia Province]. *Przedsiębiorczość-Edukacja* 3:24–35. doi: 10.24917/20833296.3.3

Zagórowska, A., R. Rauziński, and J. Słodczyk. 2020. *Procesy i struktury demograficzne na Śląsku po II wojnie światowej (województwa dolnośląskie, opolskie i śląskie)* [Demographic processes and structures in Silesia after World War II (Lower Silesian, Opole and Silesia provinces)]. Opole: Wydawnictwo Uniwersytetu Opolskiego.

Section C

Uncertainties and Risks in the Global Economy

8 Risk Narratives and Reconfigurations of Global Production Networks – The Case of the German Medical Technology Industry

Thomas Neise

8.1 Introduction

Since the 1990s, production processes in the globalization era have been fundamentally restructured through offshoring and outsourcing to reduce costs and increase the flexibility of production (Dicken 2014). As a result, the organization of business activities within global production networks has intensified and manifested itself over decades. Geopolitical dissonance, such as Brexit, Russia's invasion of Ukraine, as well as trade conflicts, climate change, and the COVID-19 pandemic, has exposed weaknesses and risks of these spatio-organizational complexities in many industries (Lund et al. 2020; Witt et al. 2021; Yeung 2023).

Against this background, debates emerged calling for a rethinking of the existing structures of the global economy to reduce the fragility and complex interconnectedness of globalized industries. This debate has been focused particularly on critical products (e.g. electronics, semiconductors, or pharmaceutical and medical goods), where disrupted supply chains can become a major risk for societies and economies worldwide (Curran and Eckhardt 2021; Miller et al. 2021; Raza et al. 2021). Currently, it seems that a new era of regionalization, or "de-risking", is shaping the thinking on the spatial organization of business activities (Enderwick and Buckley 2020; Gong et al. 2022). Consequently, the Global North, for instance, the European Union and the United States, which rely heavily on supply from East and Southeast Asia, have launched initiatives to reshore critical goods, such as pharmaceuticals and medical goods (Raza et al. 2021). Especially since the COVID-19 pandemic, global production networks have been perceived as risky and less resilient. The events in past years have shown how risk assessment can be shaped by single events or actors, and that their assessments are not always purely economically rational, but rather influenced by social processes that lead to a "labeling" of potentially risky situations.

Research in economic geography and related disciplines has begun to contribute to the debate on the impact of multiple crises and risks on (de-)globalization processes. Recently, Yeung (2023, p. 674) has pointed out that understanding risk is especially crucial for economic geographers in analyzing

DOI: 10.4324/9781003457626-11

the current and future "remaking of the global economy". In the context of global value chains or global production network studies, there is already substantial work on risk (e.g. Bryson and Vanchan 2020; Völlers et al. 2023; Schwabe 2020). However, a clear conceptualization of risk has long been neglected. Coe and Yeung (2015) were among the first to conceptually incorporate risk into their proposal for the so-called GPN 2.0 framework.[1] Their conceptual understanding of risk, however, is based largely on an objective understanding of risk which under-conceptualizes the subjective characteristic of risk. Völlers et al. (2023) have recently developed an analytical framework to address this shortcoming. Their performative risk narrative (PRN) framework sheds light on the subjective perceptions and collectively shared risk expectations of actors in a global production network. The overall aim of the framework is "to explore *where* risk is produced and *how* risk perceptions and expectations of organizational decision-makers are becoming causally relevant in global production networks" (Völlers et al. 2023, 1840, italics in the original).

Based on these conceptual ideas, this chapter will explore how PRNs shape the configuration of medical technology global production networks. The medical technology industry was chosen as a case study because of its importance during the COVID-19 pandemic, as its products, such as medical face masks and ventilators, were critical in combating the public health effects of the pandemic. Companies had to deal with a demand surge that could not be met by the firms, as the complex production processes and the supply of raw materials were disrupted (Butollo and Staritz 2022; Gereffi et al. 2022; Khan et al. 2021; Miller et al. 2021). Moreover, the industry was designated as a critical sector for health and security in Europe, where high dependence on global imports, and the resulting risks of supply disruptions should be minimized, for example, by reshoring production to Europe and expanding national production capacities (Curran and Eckhardt 2021; Butollo and Staritz 2022; Gereffi et al. 2022; Raza et al. 2021).

The chapter proceeds as follows: The next section summarizes the literature on GPN risk with a particular focus on PRNs. The third section presents the methodological approach and the medical technology industry as a case study. The fourth section discusses how PRNs shape the organizational and spatial configuration of medical technology global production networks. The fifth section shows how risk narratives within the medical technology global production networks influenced the response to supply chain disruptions in the wake of the COVID-19 pandemic. The final section concludes with the main findings of the case study.

8.2 Impact of Risk Narratives on Global Production Networks

8.2.1 Risk in Global Production Networks

The concept of GPN is one of the dominant approaches in economic geography to understand spatial disparities in a globalized economy. Global production networks are defined as "the globally organised nexus of interconnected functions and operations by firms and non-firm institutions through which goods and services are produced and distributed" (Coe et al. 2004, 471).

For a considerable time, risk has been largely unrecognized in the GPN literature. In the past decade, however, the phenomenon of risk has received increasing attention. Primary work to include risk in the debate of GPN has been done by Coe and Yeung (2015). Within their GPN 2.0 framework, Coe and Yeung (2015) include risk as one analytical category, next to the analytical categories of market imperatives, financialization, and costs. The authors argue that risk is a central dimension that influences the architecture of global production networks. Due to their increasingly complex spatial and organizational architecture, global production networks are exposed to greater risks (e.g. supply chain disruptions). At the same time, by reconfiguring their spatial and/or organizational structure, global production networks can be viable platforms for mitigating risk (Coe 2021).

According to Coe and Yeung (2015; see also Coe 2021), four genuine characteristics of GPN risk are relevant for studying risk through a GPN lens. First, through the global production network itself, risks can spread and reinforce themselves across various actors and territories (see Franz 2010; Gibson and Warren 2016; Coe 2021). Second, GPN risks are a collective dynamic, so risk mitigation usually requires a collaborative approach involving all actors within the GPN (Geenen 2018; Coe 2021; Coe and Yeung 2015). Third, risk mitigation reflects the (asymmetric) power relations in a global production network, that can result in lead firms transferring risk to other actors in the global production network (Blažek 2016; Coe 2021; Coe and Yeung 2015; Lanari et al. 2021). Fourth, particular risk manifestations are industry-specific resulting in different risk exposures (Coe 2021; Haraguchi et al. 2023; Neise et al. 2023).

Despite their valuable contribution including risk in the debate of GPN studies, several authors have challenged the understanding of risk by Coe and Yeung (2015). The main criticism by scholars, such as Bryson and Vanchan (2020) and Völlers et al. (2023), is the objective understanding of risk, inspired by the dichotomy of risk and uncertainty by Frank Knight (1921): "[O]nce [...] unknown outcomes can be predictable subject to probability distributions, they are no longer uncertain conditions [...] they can be measured as risk" (Coe and Yeung 2015, 110). This objective characterization of risk underemphasizes the actors' perspective, which is one of the focal points of the GPN 2.0 approach. Thus, the individual assessment of dynamics influencing the organization of global production networks remains barely considered. Some GPN scholars (e.g. Neilson et al. 2018; Völlers et al. 2023) have already pleaded for a more actor-centric approach that reflects the varying practices based on the context-dependent perception of events. This work is paramount as the diverse actors within a global production network do not make their decisions solely based on rational considerations of costs (i.e. risks) and benefits (i.e. rewards). Their risk assessment might be guided by social or ecological rather than pure cost minimization or profit maximization goals. Also, Vanchan et al. (2018)

stress that the decision-making of company managers is more likely to be rationally bounded and influenced by social, cultural, political, and economic circumstances.

8.2.2 Toward a PRN Perspective

Due to the shortcomings of an objective risk characterization, Völlers et al. (2023) advocate for a social constructive understanding of risk that is linked to the spatial-relational context-environment of decision-makers, to specify the explanatory nature of GPN risk. From a social constructivist perspective, GPN risk is no objective term, but is produced and constituted by the subjective perception and interpretation processes of actors. Thus, risk can be interpreted as a mode of fictional expectations (Beckert 2013) that actors use as narratives or cognitive frameworks to interpret real-world phenomena and to develop guidance for future decisions (see Clark 2022; Fuller 2023). As a future-oriented cognitive concept, risk creates socially constructed realities that give rise to actions characterized by subjective perceptions and collectively shared expectations (see Christiansen 2021; Fuller 2023; Geenen 2018). The prerequisite for understanding the management of risk in global production networks is therefore an examination of the process of perception, interpretation, and attribution of meaning and sense-making (Völlers et al. 2023).

To deconstruct how risk influences strategic decision-making within global production networks, the analytical framework of PRN by Völlers et al. (2023) aims to capture simultaneously the actor-level (i.e. decision-makers) and the structural-level (i.e. the context environment). This differentiation should help in understanding where risk is produced and how risk perceptions and expectations influence the actors' strategies and related (re-)configuration processes of global production networks. At the actor level, Völlers et al. (2023) stress that risk is a complex interplay of subjective perceptions, collective shared risk expectations, and PRNs (pre-)determining strategic decision-making within global production networks (see Figure 8.1).

Starting from a single decision-makers' perspective within a global production network, Völlers et al. (2023) stress the importance of individual risk

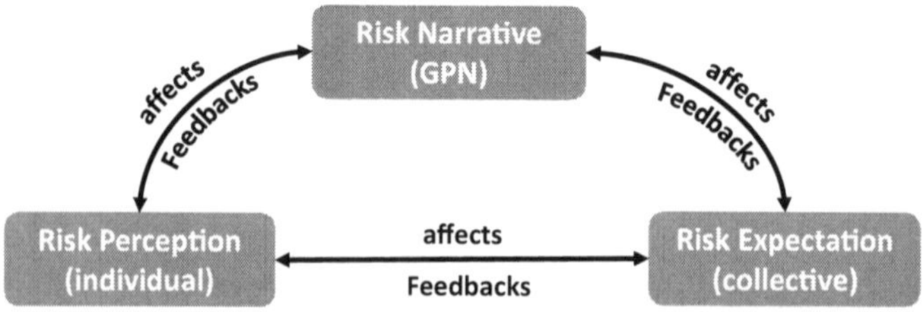

Figure 8.1 Interactions of subjective risk perception affecting collectively shared risk expectations and subsequent risk narratives. Own design based on Völlers et al. 2023.

perception. According to Aven and Renn (2010), situations are perceived as a risk if something that people value is at stake. These subjective, non-fixed interpretations are processed by individuals, and their evaluation is further influenced by personal and social-cultural beliefs and attitudes (Boholm and Corvellec 2011; Bryson and Vanchan 2020; Schweizer 2019). That means decision-makers consider how others perceive risk and respond to them by reflecting their risk perception (Douglas and Wildavsky 1983; Kasperson et al. 1988; Renn and Rohrmann 2000). As decision-makers are usually embedded in organizations, for example, a company, the framework by Völlers et al. (2023) further emphasizes collective risk expectations that emerge from a negotiation process and power relationships within an organization. These risk expectations are shaped by the decision-making environment and the situational socio-spatial context (Beckert 2013; Christiansen 2021; Fuller 2023). As a result, risk perception and risk expectations form – according to Völlers et al. (2023) – the causal logic of risk narratives within a global production network. Risk narratives become performative as they shape the strategies of all relevant decision-makers within a global production network. Moreover, they can also influence risk perception and expectations beyond the nodes of global production networks (e.g. Geenen 2018).

Besides the social constructivist understanding of risk, there are two other conceptual components of the PRN perspective: relationality and multi-scalarity. Abundant literature (e.g. Everts and Müller 2020; Fuller 2023; Müller-Mahn et al. 2018) has stressed that subjective risk perception is influenced by the socio-spatial context in which an actor is embedded. That means that, for instance, varying countries of origin, the related ownership structure, corporate culture, and political systems, shape risk perceptions and expectations (Coe 2021; Völlers et al. 2023). To uncover the spatial linkages of how risk emerges and affects the actor-based strategic formation of global production networks, the embeddedness approach by Hess (2004), with its dimensions of territorial, societal, and network embeddedness, is incorporated in the PRN perspective. Moreover, risk narratives in global production networks are driven by endogenous and exogenous dynamics, unfolding in multi-scalar interactions within organizations, between organizations, and external to organizations (see Bryson and Vanchan 2020). To conceptualize these multi-scalar interactions Völlers et al. (2023), build on the idea of cognitive context levels, which represent the decision-making environments that embody cognitive frames and sense-making logics affecting the construction of risk by decision-makers. A distinction is made between three context levels: internal (intra-organizational), network (inter-organizational), and external (extra-network) context environments. At the analytical level of the intra-organizational context environment, collective processes of meaning creation emerge to define what is considered a risk for a particular organization. The analytical level of the inter-organizational context environment focuses on the development of actor-specific risk expectations, against the background of socio-spatial network connections among GPN actors. The extra-network context environment incorporates the emergence of risk expectations external to a global production network, emphasizing the significance of the sociocultural and political context in which GPN decision-makers are situated. In the end, PRNs have a genuine geographical context, as they can be transmitted or amplified across intra-, inter-, or extra-networks (Bryson and Vanchan 2020; Coe 2021; Völlers et al. 2023). As a result, it is postulated that global production networks consist of place- and/or industry-specific risk narratives that influence the strategic decision-making of all GPN actors.

8.3 Case Study and Methods

8.3.1 Methodological Approach

To examine how PRNs influence the configuration of medical technology global production networks, the empirical research used a qualitative approach, including a structured content analysis (see Mayring and Fenzl 2019) of expert interviews and material from press and firm websites.

Between March and June 2021, eleven semi-structured interviews with medical technology companies located in Germany and two interviews with experts from business associations were conducted. The interviewees are managing directors and supply chain managers of the respective companies. All companies interviewed can be classified as medium-sized, with fewer than 250 employees. Respondents were asked about the impact of disrupted supply chains, prior risk monitoring (i.e. including risk perception), and the companies' subsequent risk mitigation strategies during the COVID-19 pandemic. Two interviews with sector experts, and a supplementary document analysis of news articles and policy documents, have been used to further understand the context environment within the medical technology sector.

All interviews were conducted in German and translated into English for this chapter. Each quote was fully anonymized in accordance with the confidentiality agreement with the interviewees.

8.3.2 German Medical Technology Sector

The medical technology industry is characterized by its broad product portfolio that includes standard mass products, such as facial masks and scalpels, as well as highly complex low-volume products, including X-ray devices, and in vitro diagnostics. All products have in common that they are intended for human use (Raza et al. 2021).

The German medical technology sector is a world leader due to its high degree of specialization and its high-quality products. The sector is dominated by medium-sized companies with less than 250 employees (Spectaris 2022, n.p.). Typically, the companies specialize in one niche segment of the broad product variety, in which they are often world market leaders. Large companies, such as B. Braun or Siemens Healthineers, which offer various product groups of medical devices, are exceptions.

The German medical technology sector is on a continuous growth path. Unlike other industries, growth has continued in the years of the COVID-19 pandemic. In the period from 2012 to 2021, sales by German medical technology companies grew by an average of 4.2% per year; with total sales of €36,4 billion in 2021. The main growth drivers are foreign markets, which account for around two-thirds of total sales (Spectaris 2022, 10). The total medical technology sector employs approximately 155,000 people directly (Spectaris 2022, 10).

During the COVID-19 pandemic, the medical technology sector raised public attention worldwide as it manufactures critical products such as personal protective equipment and ventilators. Especially at the beginning of the pandemic, supplies were disrupted as countries restricted exports, for example, China for masks and Malaysia for gloves (Gereffi et al. 2022). Companies producing related products to combat the COVID-19 infection, such as ventilators, diagnostic materials, and devices, benefitted from a surge in demand for their products. For example, the CEO of a company that manufactures ventilators had a mixed assessment of the impact of the COVID-19 pandemic on his company:

> I always compare [the COVID-19 pandemic] to a tsunami. [A] Tsunami in the positive as well as in the negative sense. On the positive side, of course, the number of inquiries multiplied enormously. [...] On the other hand, supply chains [...] have been massively disrupted, with missing deliveries or significantly increased delivery times of materials that we need to process our products in manufacturing.
>
> (interview with CEO of ventilator producer, 24.03.2021; own translation)

In contrast, companies that manufacture products or devices unrelated to the COVID-19 pandemic reported lower demand for their products.

8.3.3 Medical Technology Global Production Network

Global production networks in the medical technology sector consist of complex supplier networks and a significant influence by intermediaries, such as logistic and distribution firms, as well as certification agencies. Figure 8.2 illustrates the general structure of a global production network in the medical technology industry that exemplifies the network of medium-sized German medical technology firms. The medical technology company is the lead firm within the global production network, as it manages product design as well as research and development, generates and preserves know-how, handles the certification process, and distributes the product or services under its own brand. Moreover, it controls and coordinates the supplier network within the industry (see Coe 2021).

As numerous different components are combined in the production of medical technology products, companies source their components from many suppliers, usually worldwide. For instance, the German company Dräger, one of the world leaders in producing ventilators, mentioned that their products

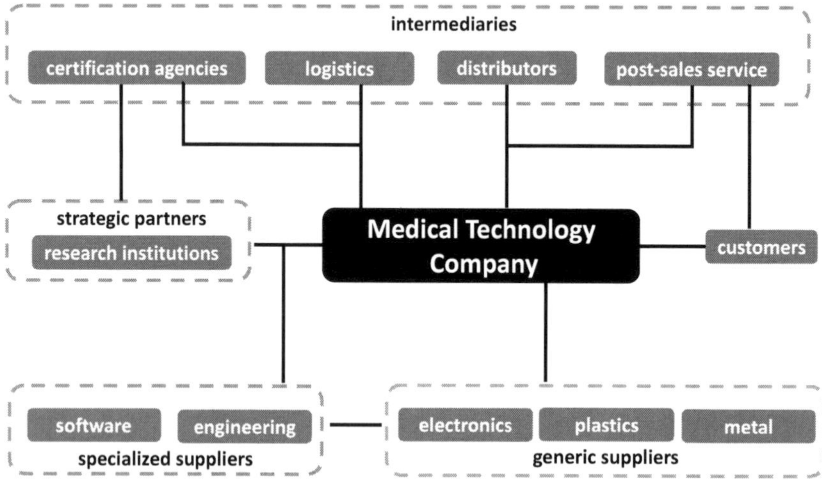

Figure 8.2 Global production network of a German medical technology company. draft: author.

consist of about 400 components, and they have about 85 suppliers (Dräger 2021, 61). Electronics, metal processing, and plastics producers are important generic suppliers, from which the medical technology companies source parts for their standard products. As a niche producer with predominately low-volume products, medical technology companies only purchase a small number of products from large electronic, metal processing, or plastic producers that usually supply high-volume sectors, such as the automobile or computer industry. Many of these suppliers are based in East and Southeast Asia. A close relationship exists with specialized suppliers, particularly software development and mechanical engineering firms that develop bespoke products for medical technology firms. These specialized firms are often located in Europe.

Certification agencies, as a regulatory intermediary, play a very influential role in organizing the production network of medical technology companies. However, their influence in the organization of the global production network is only implicitly expressed. The certification bodies do not determine the selection of suppliers, but the medical technology companies base their selection on national or multilateral legal requirements. The certification bodies influence the architecture of the global production network, as they evaluate which products from suppliers are allowed to be assembled into a medical device, and whether the medical technology products are safe. Their role is based on legal requirements, which mandate that every product and all components, must be certified to ensure that they do not cause harm to humans. These certification procedures are based on laws that have different requirements in each country or world regions. In the European Union, for example, the regulations (EU) 2017/745 on medical devices and (EU) 2017/746 on in vitro diagnostic medical devices, became effective in 2021, with transitional arrangements for already proven products until the end of 2028. Among other things, the regulations require medical technology companies to implement a risk management process for the entire life cycle of the device, as described in the ISO 14971 standard (Article 2) (VDE Health 2023). Companies must also designate a responsible person for compliance (Article 15), and each implantable device must provide an implant ID card for patients (Article 18). In addition, a former risk classification has been revised that categorizes each medical device into one of four risk categories. These categories distinguish between the severity of the potential damage that a fault or failure of the device can cause (BfArM 2023; VDE Health 2023).

8.4 PRNs in Medical Technology Global Production Networks

8.4.1 Safety Orientation as a PRN

The medical technology industry is largely characterized by its safety orientation. Since medical technology products are often used directly on humans, the products and their components are subject to strict quality and safety specifications. Patient safety is the top priority; therefore, delivery, reliability, and high quality are considered fundamental. As a result, the industry is subject to a risk narrative that stipulates that no human being may be harmed by the products. The certification bodies are the starting point for this narrative, as they specify standards and test specifications in the interests of the patient. The risk expectations, which represent the national or multilateral assessment of harm reduction by medical devices, have been formalized in legal requirements. Based on these regulations, which differ across the main global markets (e.g. European Union, United States, or China), certification bodies and companies, are required to manufacture safe medical devices.

An expert from a business association highlighted the importance of an internal and external risk assessment to reduce the potential threat to patients: "In the field of the medical technology sector, there is always an assessment of the risk to the patient [...] This then determines all the processes that take place in product development through to approval" (Interview with an expert from a business association, 05.05.2021, own translation). This demonstrates that the risk narrative of patient safety is essential for all business processes in a medical technology global production network. The companies are also respecting the general risk expectation of patient safety. Thus, this risk expectation is incorporated into the production strategies of the companies and manifests their quality control:

> It is important that the products we purchase, or medical products in general that we manufacture, that it is clear that they work, that they are certified, that everything is documented, that the processes are validated. We are not doing it for the TÜV [private certification service] or another

> service. No, this is important because we all want safe medical devices on the market. [...] We do this for ourselves and our patients.
>
> (interview with supply chain manager of a tracheostomy tube producer, 03.05.2021; own translation)

This statement also shows that medical technology manufacturers have the expectation that their products should never pose a risk to people. Interviews with company representatives made clear that the companies attach great importance to reducing the risk to patients, and often place higher safety requirements on their products than the legal requirements. These company-specific expectations arise within the intra-firm networks, where the intra-organizational contextual environment is significant. Based on the subjective risk perceptions of decision-makers in various departments of each medical technology company, collective risk expectations of the manufacturer are created, which ultimately symbolize the narrative of patient safety. This arises in internal company negotiation processes, for example, between product development, risk management, and supply chain management.

Nevertheless, the certification procedures are time-consuming and costly for companies. The newly introduced European Medical Device Regulation (MDR) in particular, is perceived by some as too bureaucratic and excessive. The MDR aims to improve patient safety through stricter requirements for market approval of medical devices such as pacemakers, implants, or prostheses. Thus, companies are required to re-certify already approved medical devices. For instance, the president of the Association of the Scientific Medical Societies in Germany (AWMF 2022, n.p.) criticized in a press statement that "[...] the current regulations of the MDR pose major hurdles for the approval of medical devices, which can lead to gaps in supply and, in the long term, to a weakening of science in Germany [...] we want to keep the implementation of the MDR practicable".

It shows that decision-makers in the medical technology sector are challenging the prevailing risk expectation, which exceeds the company's risk expectation. At the same time, the companies comply with the regulations and try to adapt their organizational structure in accordance with the MDR.

> You have to be aware that such processes simply take a long time, and in the end, you have to set yourself up internally in such a way that you can handle them well [...] and you need to know one or two shortcuts without putting patients at risk.
>
> (interview with supply chain manager of a tracheostomy tube producer, 26.04.2021; own translation)

8.4.2 Suppliers' Risk Perception and Risk Expectations

The combination of the risk expectations of the medical technology companies and certification bodies gives rise to a PRN of patient safety. This PRN is transferred to the suppliers of the medical technology manufacturers, as they are also obliged to certify machines and products. Compliance with the regulations is a fundamental basis for business relationships between suppliers and medical technology manufacturers. Therefore, the selection of suppliers is usually part of the certification of a new product; thus, medical technology companies select their suppliers meticulously. To avoid substitution of suppliers, and to minimize quality problems and associated risks in certification, manufacturers enter into long-term contractual relationships, which are often also close personal contacts, as the interview quote shows:

> How secure is their material supply? Are there any changes to the basic material? We need to know all this in advance and make appropriate supplier contracts, but [the] paper is patient. You can write anything into a contract. That is why we tend to take the supplier in our arms, get them close to us, to make sure that production and delivery work.
>
> (interview with operations manager of a tracheostomy tube producer, 26.04.2021; own translation)

German medical technology manufacturers prefer suppliers that are located nearby and are familiar with the requirements. The companies interviewed mentioned that geographical and relational proximity helps to identify potential supply problems and to supervise the quality. This consideration is also reflected in the geographical distribution of the German medical technology industry. The industry is spatially dispersed throughout Germany with regional clusters of specialized production, such as Tuttlingen (in the south of Germany) for surgical devices, or northern Germany for ventilators (Roitzsch et al. 2021). It can therefore be concluded that the embeddedness of suppliers in the inter-organizational network is an important interface to clarify the manufacturer's risk expectations on the one hand and to support and monitor compliance on the other.

However, costs also play an important role in selecting suppliers. Spatial proximity is not the sole decisive factor, but firms weigh the costs and the spatial proximity together, as well as the reliability of supply. Furthermore, the medical technology companies need to accept the spatial concentration of the production of generic electronic, plastic, and metal parts that takes place to a large extent in East and Southeast Asia. The producers are powerful actors in the medical technology global production network. From the perspective of Asian manufacturers, medical technology companies are often marginal customers, as they only order very small quantities compared to their major customers in the automotive or electrical engineering sectors. As a result, the production and delivery processes are geared toward the prevailing just-in-time production of their major customers, who contractually guarantee a high purchase volume. One operations manager interviewed gave the example of his weak negotiating position with their generic suppliers, which reveals his corresponding risk expectation:

> We also get parts from China. We are super unhappy with that because if there is a problem and you call them, they do not answer the phone. They just do not care. If you write them an email, if it is an unpleasant issue, they do not answer either. They play dead.
>
> (interview with operations manager of a tracheostomy tube producer, 26.04.2021; own translation)

Reliability of supply is a key determinant of the risk expectation in the medical technology sector. For each new product, the medical technology companies consider a safety stock of the materials supplied. Based on the internal supply risk management strategy and discussions with potential suppliers, the companies decide whether they build their own safety stock or require suppliers to hold a safety stock.

> And we here at [company] have always paid more attention to safety. We have always given priority to delivery reliability over low stock volumes. The ability to deliver is more important to us than saving interest or cash flow or anything else in the short term. That was never our focus!
>
> (interview with plant manager of company producing plastic-based medical products, 10.06.2021; own translation)

8.5 Responding to Supply Chain Disruptions in the Wake of the COVID-19 Pandemic

The characteristics of the risk narrative, and the corresponding strategies of medical technology companies, also had a considerable influence on their response to supply chain disruptions during the COVID-19 pandemic. All the companies surveyed perceived the consequences of the COVID-19 pandemic as a risk to their supply chains and production. The restricted global trade has created difficulties in the procurement of primary products. Suppliers were unable to fulfill their supply contracts because production was suspended or reduced, and transport options were limited, resulting in longer transport times.

> We source components from Taiwan and China. Customs has first closed the borders. Especially when [the components are used] for medical equipment and then heard ventilation, then you knew that gets trouble because they also wanted to keep the products first in their own country.
>
> (interview CEO of ventilator company, 1.05.2021, own translation)

The companies were aware of their susceptibility to supply chain risks, as they must rely on a limited number of often specialized suppliers. Experiences of supply chain disruptions during the world financial crisis in 2007/2008, the Tohoku Earthquake in Japan in 2011, and Brexit, sensitized the companies and influenced their risk expectations, which also determined their response strategies with the beginning of the COVID-19 pandemic. As many companies are active in China, one of the largest medical technology markets, they were immediately aware of the spread of the COVID-19 virus. As a result, they took measures for their own business, and a possible global spread of the virus was taken into account in ordering processes and production planning.

> Yes, we already saw the situation at the beginning of the year, because we are also active in China, and we saw a significant increase in business in China as early as January. We then responded relatively promptly in various ways with what we could
> (interview with CEO of ventilator company, 24.03.2021, own translation)

The companies' risk mitigation strategies included stockpiling critical raw materials and intermediate products for production for a few months to a year. Companies were therefore able to cope relatively well with the first global lockdown with closed borders as they were able to fall back on their stocks. In addition, *ad hoc* measures flanked the company's risk mitigation. However, meeting the surge in demand was only possible to a certain extent. So, the companies extended the production time, by adding another shift, integrating administrative staff into the production, or borrowing staff from other companies which reduced their production. Satisfying the high demand with a major expansion of production was hardly possible, as new production lines required months of certification, and preliminary products were difficult to obtain.

With the prolonging of the COVID-19 pandemic and the restart of many manufacturing companies, such as the automotive sector, the medical technology industry faced increasing difficulties in sourcing preliminary products, such as electronic equipment and hardware chips.

> *[…] we do have difficulties in the procurement of electronic components. Not only the automotive industry is affected, but we also have massive problems with unstable delivery dates or the non-availability of spare parts. (interview with CEO of a diagnostic company, 18.06.2021, own translation).*

Although the companies attempted to mitigate their susceptibility to supply chain disruptions by long-term supply contracts, they suffered more and more with being unable to deliver. Concerning their risk expectations, this dilemma was drastic for the medical technology companies, as the inability to deliver ultimately means that people's health is at stake.

At the same time, the PRN of the medical technology sector concerning detailed certification processes, made it nearly impossible to add further suppliers. The medical technology companies were heavily dependent on their existing suppliers being able to deliver the products. Furthermore, it was hardly possible to ramp up production capacities, as machines and suppliers were not certified.

> Creating new production capacities is of no use if the procurement channels are not clearly structured and they use the network to get the materials

> they need to produce devices. […] normally an electronic part is delivered in two weeks. Suddenly the supplier tells you four or five months. Then I start to struggle and that's what happened to everyone […] at the same time people are knocking on your door and saying they want your devices.
> (interview CEO of a ventilator company, 1.05.2021, own translation)

To a certain extent, the medical technology companies could manage the supply chain disruption, as they proactively considered disruptions when developing a new product and selecting suppliers. This risk expectation is reflected especially in their risk-averse sourcing strategy of critical components. The companies discuss safety stock levels among the departments and with the potential suppliers.

> *Is the supplier able to hold a safety stock or do we hold the safety stock? The issues are considered in advance when a product or an individual component is newly developed. (interview supply chain manager of products for cardiovascular imaging, 10.06.2021, own translation).*

Other companies are shifting the delivery reliability to their suppliers in East or Southeast Asia. The medical technology companies set up consignment warehouses in Germany or directly on-site, so that the suppliers can stock up on materials independently. However, this strategy is only possible for large-volume materials, and for suppliers that have other customers in Germany.

Another common strategy is dual sourcing, where at least one second supplier is close by to act as a back-up. In this case, the companies accept a higher price as the supplier is typically only designated to supply 20% of the demand.

> If we have the opportunity and we know that something is coming from Asia or from some crisis region or it is a critical component process where we need longer for certification, we naturally always try to use a local secondary source […]. The smaller quantity may be more expensive, but the supplier is on standby in case any shipment fails.
> (interview supply chain manager of products for cardiovascular imaging, 10.06.2021, own translation)

Another interviewee, however, mentioned the limitations of relying on a second (local) source, as the medical technology companies typically source low volumes: "And what do I tell supplier B? If the worst comes to the worst, you can step in. Who really has a major economic interest in this?" (interview with supply chain manager of a tracheostomy tube producer, 26.04.2021, own translation). As a result, companies are trying to order larger quantities of components that will last for months or up to a year. In addition, companies' risk expectations lead them to operate a risk-averse, in-house risk management. Companies with several production lines certify several machines in case one production line falls out. One interviewee, from a company that also has a production unit in the United States, explained that the production line for the US market is also certified for the European market, and vice versa. In normal times, production is separated as different legal requirements exist between the European Union and the United States; however, in an emergency, the companies take the financial and organizational resources to certify the production lines for both markets, in order to have a replacement in case of failure.

The difficulties in responding to supply chain disruptions reveal that the medical technology companies are constrained in their reconfiguration options. On the one hand, the companies are inhibited by the risk expectations of their global suppliers, that orient their production procedure (i.e. just-in-time production) toward their main customers from the automotive and electronics sector. On the other hand, their reconfiguration options are shaped by the risk expectations of certification agencies that represent the (multi-)national legal regulations. Long-term certification processes that should ensure the quality and safety of medical devices make it hard to respond flexibly to supply chain disruptions.

The COVID-19 pandemic has also shown how the extra-network contextual environment could change the spatial architecture of medical technology global production networks. Given the supply bottlenecks and the production of critical components outside Europe, the debate arose about bringing production steps back to Europe. The risk narrative of patient safety is subjected to a further component that was introduced by the society and politics, namely the risk expectation that patients are at risk if the products cannot be supplied in sufficient quantities. A study by the European Commission (Raza et al. 2021), which analyzed the possibilities of reshoring the production of medical devices to Europe, emphasizes that reshoring medical devices is primarily motivated by political reasons (e.g. enhancing regional competitiveness and securing the supply of critical goods) and less by rational economic considerations of the companies. The study also points out that a large-scale relocation of medical technology companies is not necessary, as Europe already has large production capacities. More problematic is the forced single sourcing, as for many components only a limited number of suppliers are available who are not exclusively based in Europe (Raza et al. 2021). Thus, the authors advocate implementing policies that enhance supply resilience and encourage the stockpiling of components. These are risk management strategies which, as the interviews showed, are already being put into effect in medical technology companies.

8.6 Concluding Discussion

Today's global production networks are at risk due to the increasing complexity of global sourcing that becomes constrained due to trade wars, pandemics, and extreme weather events. Although GPN scholars have turned their attention to the impacts of risks on the spatial organization of global production networks, the specific risk-related effects remain vague. Consequently, recent literature has claimed that GPN risks should be seen as socially constructed by influential actors within a global production network.

Using the proposal of a PRN perspective within GPN literature, this chapter used the example of the German medical technology industry to examine how risk narratives determine the reconfiguration strategies to reduce the risk of disrupted supply chains. The case study shows that medical technology's PRN of patient safety influences the spatial organization of their global production networks. Although cost leadership is part of the consideration in organizing the production networks, reliability of supply and trust are critical to reflect the risk narrative, and the own risk expectations of the medical technology companies. However, medical technology companies face difficulties in passing on the risk narrative to their generic suppliers operating from East or Southeast Asia. These companies tend to orient themselves toward their risk expectations and the risk narratives of the automotive and electronics industries. This became particularly clear during the supply chain disruptions caused by the COVID-19 pandemic. The diverse risk expectations contribute to the fact that medical technology companies have a limited ability to reconfigure their supply network in such a way that a reliable supply is guaranteed, and they can respond flexibly to supply interruptions.

Regarding the contemporary debates on regionalization and reshoring, two key implications can be drawn from the case study. First, reconfiguration strategies depend, not only on the risk expectations within a business sector, but also on the risk expectations of global suppliers, which are also embedded in the risk narratives of other, more powerful, global production networks. For example, moving all components of medical technology global production networks to Europe remains a major challenge. Second, it becomes clear that risk narratives have a dynamic character, and that major events with a meaningful impact on social or political contexts, can change their performative nature. Due to the overlapping crises, in particular the COVID-19 pandemic and trade wars, supply reliability has become an increasingly dominant rationale and an additional risk expectation for politicians and medical technology companies. Supply reliability is not only an economic risk factor in medical technology, but also a further element of patient safety. Although the reliability of supply, which is expressed in the context of the choice of country of origin ("friend-shoring"), has not yet been formalized in the regulations and laws on patient safety. However, it has become a critical element that is implicitly considered by medical technology companies and discussed by policymakers who refer to the medical technology sector as a critical industry.

In conclusion, the case study has shown that it is worth focusing on the subjective understanding of risk, as it helps to identify the different decision-makers' perceptions and organizational expectations that determine the prevailing PRN in global production networks. These PRNs provide an explanation of the spatial architecture of the global production network of medical technology and its difficulties in risk mitigation through reconfiguration. In this context, it would be promising to devote future research to the underlying power relations that manifest the performativity of risk narratives that determine risk mitigation strategies within a global production network.

The divergent risk perceptions and expectations of all GPN actors involved should be examined. Furthermore, it is worth analyzing the spatio-temporal emergence of sector-specific PRN, especially in times of multiple crises, to decipher the driving risks that influence the spatial organization of global production networks and potential risk-related shifts.

Note

1 The GPN 2.0 framework is a further conceptual development of the GPN approach that stresses the dynamic character of global production networks. The acronym 'GPN' is used to refer to the GPN approach, and "global production networks" means the empirical phenomenon (see Coe and Yeung 2015).

References

Aven, T., and O. Renn. 2010. *Risk Management and Governance: Concepts, Guidelines and Applications*. Berlin: Springer. doi: 10.1007/978-3-642-13926-0

AWMF. 2022. Kritik an europäischer Medizinprodukteverordnung. Praktikabilität nur durch gezielte Zusatzmaßnahmen – eine Gefahr für die Innovationskraft [Criticism of the European Medical Device Regulation. Practicability only through targeted additional measures - a threat to innovative strength]. Accessed January 8, 2024. https://www.awmf.org/fileadmin/user_upload/dateien/presse/PM_AWMF_Positionspapier_MDR_Dez_2022_F.pdf

BfArM. 2023. Abgrenzung und Klassifizierung [Delimitation and classification]. Accessed December 11, 2023. https://www.bfarm.de/DE/Medizinprodukte/Aufgaben/Abgrenzung-und-Klassifizierung/_node.html

Beckert, J. 2013. Imagined futures: Fictional expectations in the economy. *Theory and Society* 42 (3):219–40. doi: 10.1007/s11186-013-9191-2

Blažek, J. 2016. Towards a typology of repositioning strategies of GVC/GPN suppliers: the case of functional upgrading and downgrading. *Journal of Economic Geography* 16 (4): 849–69. doi: 10.1093/jeg/lbv044

Boholm, Å., and H. Corvellec. 2011. A relational theory of risk. *Journal of Risk Research* 14(2):175–190. doi: 10.1080/13669877.2010.515313

Bryson J. R., and V. Vanchan. 2020. Covid-19 and Alternative Conceptualisations of Value and Risk in GPN Research. *Tijdschrift voor economische en sociale geografie* 111 (3):530–42. doi: 10.1111/tesg.12425

Butollo, F., and C. Staritz. 2022. Deglobalisierung, Rekonfiguration oder Business as Usual? COVID-19 und die Grenzen der Rückverlagerung globalisierter Produktion [Deglobalization, reconfiguration or business as usual? COVID-19 and the limits of the relocation of globalized production]. *Berliner Journal für Soziologie* 32 (3):393–425. doi: 10.1007/s11609-022-00479-5

Christiansen, J. 2021. Fixing fictions through blended finance: The entrepreneurial ensemble and risk interpretation in the Blue Economy. *Geoforum* 120:93–102. doi: 10.1016/j.geoforum.2021.01.013

Coe, N. M., M. Hess, H.W.-C. Yeung, P. Dicken, and J. Henderson. 2004. 'Globalizing' regional development: A global production networks perspective. *Transactions of the Institute of British Geographers* 29:468–84. doi: 10.1111/j.0020-2754.2004.00142.x

Coe, N. M. 2021. *Advanced introduction to global production networks*. Cheltenham: Edward Elgar.

Coe, N. M., and H. W.-C. Yeung. 2015. *Global production networks. Theorizing economic development in an interconnected world*. Oxford: Oxford University Press.

Curran, L., and J. Eckhardt. 2021. Why Covid-19 Will Not Lead to Major Restructuring of Global Value Chains. *Management and Organization Review* 17 (2):407–11. doi: 10.1017/mor.2021.18

Dicken, P. 2014. *Global Shift. Mapping the Changing Contours of the World Economy*. 7th ed. London: Sage.
Douglas, M., and A. Wildavsky. 1983. *Risk and Culture. An Essay on the Selection of Technological and Environmental Dangers*. Oakland: University of California Press.
Dräger 2021. Corona und die Folgen [Corona and the consequences]. Drägerheft Ausgabe 1/2021. Lübeck: Dräger.
Enderwick, P., and P. J. Buckley. 2020. Rising Regionalization: Will the Post-Covid-19 World See a Retreat from Globalization. *Transnational Corporations* 27 (2):99–112. doi:10.18356/8008753a-en
Everts, J., and K. Müller. 2020. Riskscapes, politics of scaling and climate change: towards the post-carbon society? *Cambridge Journal of Regions, Economy and Society* 13 (2):253–66. doi: 10.1093/cjres/rsaa007
Franz, M. 2010. The potential of collective power in a global production network: UNICOME and Metro Cash & Carry in India. *Erdkunde* 64(3): 281–90.
Fuller, C. 2023. Uncertainty, fictional expectations and economic agency. *Geoforum* 140:103699. doi: 10.1016/j.geoforum.2023.103699
Geenen, S. 2018. Underground Dreams. Uncertainty, Risk and Anticipation in the Gold Production Network. *Geoforum* 91:30–38. doi: 10.1016/j.geoforum.2018.02.019
Gereffi, G., P. Pananond, and T. Pedersen. 2022. Resilience Decoded: The Role of Firms, Global Value Chains, and The State in Covid-19 Medical Supplies. *California Management Review* 64 (2):46–70. doi: 10.1177/00081256211069420
Gibson, C., and A. Warren. 2016. Resource-Sensitive Global Production Networks: Reconfigured Geographies of Timber and Acoustic Guitar Manufacturing. *Economic Geography* 92(4): 430–54.
Gong H., R. Hassink, C. Foster, M. Hess, H. Garretsen. 2022. Globalisation in reverse? Reconfiguring the geographies of value chains and production networks. *Cambridge Journal of Regions Economics and Society* 15:165–81. doi: 10.1093/cjres/rsac008
Haraguchi, M., T. Neise, W. She, and M. Taniguchi. 2023. Conversion Strategy Builds Supply Chain Resilience During the Covid-19 Pandemic: A Typology and Research Directions. *Progress in Disaster Science* 17:100276. doi: 10.1016/j.pdisas.2023.100276
Hess, M. 2004. 'Spatial' relationships? Towards a re-conceptualisation of embeddedness'. *Progress in Human Geography* 28:165–86. doi: 10.1191/0309132504ph479oa
Kasperson, R. E., O. Renn, P. Slovic, H. S. Brown, J. Emel, R. Goble, J. X. Kasperson, and S. Ratick. 1988. The Social Amplification of Risk: A Conceptual Framework. *Risk Analysis* 8 (2):177–87.
Khan, S., A. Haleem, S. G. Deshmukh, and M. Javaid. 2021. Exploring the Impact of Covid-19 Pandemic on Medical Supply Chain Disruption. *Journal of Industrial Integration and Management* 6 (2):235–55. doi: 10.1142/S2424862221500147
Knight, F. H. 1921. *Risk, Uncertainty, and Profit*. Boston: Houghton Mifflin.
Lanari, N., D. Bek, J. Timms, and L. Simkin. 2021. In Whose Interests? Water Risk Mitigation Strategies Practiced by the Fruit Industry in South Africa's Western Cape. *Geoforum* 126:105–14. doi: 10.1016/j.geoforum.2021.07.025
Lund, S., J. Manyika, J. Woetzel, E. Barrball, M. Krishna, K. Alicke, M. Birshan, K. George, S. Smit, D. Swan, et al. 2020. Risk, resilience, and rebalancing in global value chains. Accessed December 12, 2023. https://www.mckinsey.com/capabilities/operations/our-insights/risk-resilience-and-rebalancing-in-global-value-chains?cid=eml-web
Mayring, P., and T. Fenzl. 2019. Qualitative Inhaltsanalyse [Qualitative content analysis]. In *Handbuch Methoden der empirischen Sozialforschung* [Handbook of methods of empirical social research], ed. N. Baur, and J. Blasius, 633–48. Wiesbaden: Springer. doi: 10.1007/978-3-658-21308-4_42

Miller, F. A., S. B. Young, M. Dobrow, and K. G. Shojania. 2021. Vulnerability of the Medical Product Supply Chain: The Wake-Up Call of Covid-19. *BMJ Quality & Safety* 30 (4):331–35. doi: 10.1136/bmjqs-2020-012133

Müller-Mahn, D., J. Everts, and C. Stephan. 2018. Riskscapes revisited - exploring the relationship between risk, space and practice. *Erdkunde* 72(3):197–213.

Neilson, J., Pritchard, B., Fold, N., and A. Dwiartama. 2018. Lead firms in the cocoa–chocolate global production network: An assessment of the deductive capabilities of GPN 2.0. *Economic Geography* 94(4): 400–24.

Neise, T., T. López, and A. F. Angga Reksa. 2023. Rethinking Labour Risk in Global Production Networks: Resilience Strategies of Cruise Ship Workers in the Wake of the Covid-19 Pandemic. *Geoforum* 145:103842. doi: 10.1016/j.geoforum.2023.103842

Raza, W., J. Grumiller, H. Grohs, J. Essletzbichler, and N. Pintar. 2021. *Post Covid-19 Value Chains: Options for Reshoring Production Back to Europe in a Globalised Economy*. Brussels: European Union.

Renn, O., and B. Rohrmann. 2000. Cross-cultural risk perception research: State and challenges. In *Cross-Cultural Risk Perception. A Survey of Empirical Studies*, ed. O. Renn and B. Rohrmann, 211–33. Dordrecht: Springer.

Roitzsch, C., T. Neise, and S. Henn. 2021. Medizintechnik in Deutschland – unscheinbar, aber weltweit an der Spitze [Medical technology in Germany – inconspicuous, but world-leading]. Nationalatlas Aktuell. Accessed December 13, 2023. https://aktuell.nationalatlas.de/wp-content/uploads/21_01_Medizintechnik.pdf

Schwabe, J. 2020. Risk and counter-strategies: The impact of electric mobility on German automotive suppliers. *Geoforum* 110:157–67. doi: 10.1016/j.geoforum.2020.02.011

Schweizer, P.-J. 2019. Systemic risks - concepts and challenges for risk governance. *Journal of Risk Research* 24 (1):78–93. doi: 10.1080/13669877.2019.1687574

Spectaris. 2022. *Die deutsche Medizintechnik-Industrie* [The German medical technology industry.]. *SPECTARIS Jahrbuch 2022/2023*. Berlin: Spectaris.

Vanchan, V., R. Mulhall, and J. Bryson. 2018. Repatriation or Reshoring of Manufacturing to the U.S. and UK: Dynamics and Global Production Networks or From Here to There and Back Again. *Growth and Change* 49 (1):97–121. doi: 10.1111/grow.12224

VDE Health. 2023. Die EU Medizinprodukteverordnung (MDR): Was ändert sich? [The EU Medical Device Regulation (MDR): What is changing?]. Accessed December 9, 2023. https://www.vde.com/topics-de/health/beratung/die-eu-medizinprodukteverordnung-mdr-was-aendert-sich

Völlers, P., T. Neise, P. Verfürth, M. Franz, F. Bücken, and K. P. Schumacher. 2023. Revisiting Risk in The Global Production Network Approach 2.0 - Towards A Performative Risk Narrative Perspective. *Environment and Planning A: Economy and Space* 55 (8):1838–58. doi:10.1177/0308518X231169288

Witt, M. A., P. P. Li, L. Välikangas, and A. Y. Lewin. 2021. De-globalization and decoupling: Game Changing Consequences? *Management and Organization Review* 17 (1):6–15. doi: 10.1017/mor.2021.9

Yeung, H. W.-C. 2023. Troubling economic geography: New directions in the post-pandemic world. *Transactions of the Institute of British Geographers* 48 (4):672–80. doi: 10.1111/tran.12633

9 Barriers and Risks in Digitalization and Industry 4.0 Process

Evidence from Carpet Industry in Gaziantep, Turkey

Nuri Yavan, Ceyda Kurtar Anlı, and Mehmet Ragıp Kalelioğlu

9.1 Introduction

Technological innovations have a significant role in the revitalization of economic growth and productivity with each industrial revolution. Advances in information and communication technology offer competitive advantages for nations, regions, and firms by increasing the flow of innovative products and services. Digitalization and Industry 4.0 technologies present various opportunities, risks, and barriers for manufacturing firms, regions, and countries, depending on their competitive position. While digitalization provides a comprehensive set of digital tools and infrastructure for innovation outputs at all levels of the business model (Isaksen et al. 2020), Industry 4.0 is an advanced technological process built on cyber-physical systems enabling the collection, sharing, and analysis of data (Götz and Jankowska 2017; Kiss and Nedelka 2020). Although still in its infancy, this process has significantly transformed manufacturing firms' productivity, quality, and competitive position, as well as their relationships with employees and customers (Lazzeretti et al. 2023; Winter 2020).

The processes and changes toward digitalization and Industry 4.0, which vary according to the type of sector, size of the firm, and place/region (Balland and Boschma 2021; Dyba, Di Maria, and Chiarvesio 2022), are fairly new research topics in economic geography (Fraske 2022; Winter 2020). In general, the studies on digitalization and Industry 4.0 in economic geography can be examined in two different strands of research. *The first strand* is the studies focusing on the role and economic effects of *digitalization* (Haefner and Sternberg 2020) in structuring value chains but paying little attention to its spatial implications. Within this research strand, the few studies handling spatial effects examine whether digitalization plays a role in increasing inequality between rural and urban areas at the national and sub-national levels and what opportunities, risks, and barriers to digitalization it may face (Haefner and Sternberg 2020; Hervás-Oliver 2021). Others try to provide empirical evidence on how digitalization as an innovation process takes place in a regional context (Isaksen, Langemyr Eriksen, and Rypestøl 2020, Isaksen et al. 2020). *The second strand* is made up of studies dealing with *Industry 4.0* (Fraske 2022). Some

DOI: 10.4324/9781003457626-12

of these studies have looked into the awareness, readiness, and adaptation of the manufacturing sector at national and regional levels (Corò and Volpe 2020; Dyba, Di Maria, and Chiarvesio 2022; Szalavetz 2019). Another group of studies has looked at clusters (Bellandi et al. 2020; Götz and Jankowska 2017; Hervas-Oliver et al. 2019), the value chain (Kinkel 2020), Industry 4.0 technologies (Lazzeretti et al. 2023) and how they contribute to knowledge production (Balland and Boschma 2021). Moreover, some studies have focused on the drivers (Corò and Volpe 2020; Horváth and Szabó 2019), barriers (Raj et al. 2020; Jankowska et al. 2022), and opportunities (Capello and Lenzi 2021; Horváth and Szabó, 2019; Schwab 2016) of the transition to Industry 4.0, as well as the risks (Frey and Osborne 2017; Völlers, Yavan, and Franz 2021; Völlers et al. 2023) arising from digitalization and their impact on the labor force.

While digitalization and Industry 4.0 create new opportunities for economic and social progress, they also bring new risks and challenges. Governments are taking various digital initiatives to identify risks and minimize the negative effects that could arise in the manufacturing sector and other areas of different scales due to digitalization. With national documents and policies such as the "*Industry and Technology Strategy 2023*" and the "*Digital Turkey Roadmap*", Turkey, like other countries, is striving to upgrade the technological level of its industrial legacies to compete at a global level by addressing deficiencies in infrastructure, finance, and governance. However, unlike the problems of developed countries, the weakness of the infrastructural base and the innovation system in Turkey and the limited availability of a qualified labor force are among the main barriers to the technological transformation (Erdil 2023). In addition, Turkey is one of the OECD countries where the existing unskilled labor force is at a high risk of losing their jobs due to digitalization (OECD 2018). Moreover, recent developments in Turkey have led to significant changes in the institutional environment, posing challenges for both firms and organizations and placing the country in a high institutional risk position (Völlers, Yavan, and Franz 2021; Yavan, Yılmaz, and Aniç 2022).

On the other hand, the challenges and problems faced by Turkey as a whole in the process of digital transformation are much deeper, both across regions of the country and at the scale of industrial districts. In this context, Gaziantep is one of the most important *industrial districts* in Turkey and has been making a significant effort in the process of digitalization and transition to Industry 4.0 in recent years. It presents an interesting case in terms of trying to digitalize its very strong but technologically fragile manufacturing industry. Accordingly, Gaziantep, which is our case study area, is both part of the global production network and a city where low-tech and labor-intensive industrial production is concentrated (Bayırbag 2010; Eraydın 2002; Oz 2004), with traditional sectors such as textiles, leather, paper, and food. As part of Turkey's digital roadmap, the awareness of Gaziantep's industry is being raised, and the technological transformation process is being accelerated through both the public sector's incentives for firms in the digitalization process and initiatives such as the Gaziantep's Chamber of Industry's "Competence and Digital Transformation

Center" (Model Factory). To this end, using the case of Gaziantep carpet manufacturing to understand the digitalization and Industry 4.0 transition process of labor-intensive low-tech industrial districts, the study attempts to identify and uncover the barriers that prevent companies from digitalization, the risks they face in this process, and the opportunities they capture. Therefore, this study contributes to understanding the digitalization processes of firms in economic geography overall by identifying potential risks that may arise with digitalization and, in particular, by illuminating the technological transformation of traditional industrial regions.

The study is divided into five sections. In the second section, following the introduction, the definition and awareness of digitalization and Industry 4.0 are assessed, along with the conceptual and theoretical framework that explains the barriers, risks, and opportunities of technological transformation. The third section describes the collected data and the techniques used to analyze it. The fourth section provides a detailed analysis of the findings obtained from the Gaziantep carpet industry, while the final section contains a discussion and conclusion that relate the main results of the findings to the literature and provide recommendations that should be made in this field.

9.2 Theoretical Background and Literature Review

9.2.1 *Digitalization and Industry 4.0*

Digitalization is defined as obtaining digital products and data by converting analog data into digital data sets, businesses integrating these products and data into their business processes, and adopting a digital structure (Kıyak 2023). Besides the fact that digitalization or digital technologies are used as a general concept in many fields, their definition also varies according to the areas of use. Digitalization, or "digital transformation" and the core technologies of Industry 4.0, seen as a revolution in the industry, are based on evolutionary accumulated knowledge generated by different techno-economic paradigms underpinning the economy and society (Isaksen et al. 2020, 132). The emergence of the *first industrial revolution* was triggered by the invention of the steam engine, which enabled the mechanization of simple tasks and the construction of railways during the 1760s; the *second industrial revolution* was triggered by the advent of electricity, the assembly line, and mass production in the 1870s; and *the third industrial revolution* was triggered by the development of semiconductors, the internet, application software, personal computers, microprocessor-based computer numerical control (CNC) technology, communication technologies, and computer networks on a global scale since the early 1970s onwards (UNIDO 2020). In this third industrial revolution period, also known as *the Computer/Digital Age or the Digital Revolution/Transformation*, programs such as enterprise resource planning (ERP), supply chain management (SCM), machine data collection (MDC), product lifecycle management (PLM), manufacturing execution systems (MES), product data management (PDM), material resource planning (MRP), and computer-aided manufacturing (CAM) increase the connection between computers used in business processes and machines used in production processes, creating new products, processes, and business models, making traditional product and sector definitions meaningless (Kıyak 2023). Another factor that is as important in digital transformation as the integration of digital technologies into production processes is the change and adaptation of organizational processes (Isaksen, Langemyr Eriksen, and Rypestøl 2020; Isaksen et al. 2020; Kıyak 2023). This is because "human and institutional assets that depend on the need to develop and reuse existing knowledge, skills, and ways of doing business" are very important for the realization of digitalization (Isaksen, Langemyr Eriksen, and Rypestøl 2020, 1465). Perhaps most important for digital transformation is the development of new competencies, and skills in the workforce, new firm competencies, new public attitudes and know-how, including changes in organizational and institutional support infrastructure (Isaksen et al. 2020, 134).

While computer-based automation brought about by digital transformation is limited to routine tasks involving activities based on specific and defined rules, the emergence of factories suitable for the production of small quantities of individualized products through the digitization of manufacturing, communication technologies, integrated technologies, and new integrated systems has become possible with Industry 4.0 (TTGV 2018). With the *fourth industrial revolution* that started in the 2010s, a process built on cyber-physical systems, called *Industry 4.0*, has begun in which machines communicate with each other, and digitalization has now reached the highest level through data sharing and data targeting the process (Haefner and Sternberg 2020). What distinguishes and differentiates this process from other techno-economic paradigms is that Industry 4.0 has forty future-determining technologies, ten of which are "radical" (Laffi and Boschma 2022, 6) and "disruptive" (TTGV 2018). This is because these disruptive technologies (Blockchain, Big Data Analytics, Artificial Intelligence (AI), Internet of Things (IoT), Neurotechnologies, Synthetic Biology, Nanomaterials, Additive Manufacturing, Micro and Nano Satellites, and Advanced Energy Storage Technologies) enable the radical progress and transition from the Industry 3.0 era of flexible automation through CNC machines to Industry 4.0 (Fromhold-Eisebith et al. 2021) by changing the production process as a whole and making it possible to connect each step of the process in an integrated way.

Compared to previous industrial revolutions, the fourth industrial revolution is characterized by a higher level of connectivity between the competitive environment, processes, and actors, potentially designed by technologies such as automation, IoT, and AI. Industry 4.0 introduces such disruptions and transformations in five key areas: new markets, new products, product-service innovation, new business models, and customer-centric innovation (Bellandi et al. 2020; De Propris and Bailey 2020). In other words, Industry 4.0 has created an entirely new production model within the factory and between firms (De Propris and Bailey 2020, 5–6). Firms are now able to redesign their business models in different ways, in particular, to deliver value to customers within a servitization framework (Hervás-Oliver 2021, 3). Although Industry 4.0 technologies that have emerged in the last 10 years are basically built on ICT-based hardware and software technologies developed within the Industry 3.0 paradigm that emerged after the 1970s, which shows that there is a significant degree of continuity between Industry 3.0 and 4.0 technologies, the distinctive (other) feature of Industry 4.0 is that while on the one hand advanced 3.0 technologies are applied in new areas where they have not been applied before (e.g., the use of IoT technologies in agriculture or the application of AI tools to legal and business services), they also have radically original and potentially disruptive innovations (Laffi and Boschma 2022, 6). Although computerized automation is at the center of Industry 3.0, this system is limited to specific tasks. In contrast, Industry 4.0 has the ability to work and functionalize in detail up to the digitalization of production, the reduction of labor-based production, and the production of small quantities of individualized products through intelligent and newly integrated systems. In particular, with the IoT, units or objects have the ability to work together or individually, depending on their purpose and environment (TTGV 2018). In short, Industry 4.0 is not only a transformation of modern business models but also a process of creative destruction that breaks old patterns and creates new ones – a transformation not only of production but also of all business cycles and value creation outside of production (Götz 2020, 55–56).

9.2.2 Barriers and Risks in Digital Transformation and Industry 4.0

During the digital transformation and transition to Industry 4.0, firms may face barriers, with different effects depending on the firms, both *internal* barriers consisting of organizational issues and problems preventing collaboration, as well as *external* barriers associated with legal issues (Jankowska et al. 2022). Differences in the impact of these barriers are strongly correlated with firm size. For example, Jankowska et al. (2022) show that in Poland, the largest adopters of disruptive technologies and the quickest problem solvers are large firms outside urban agglomerations. Similarly, the adoption of Industry 4.0 technologies is much easier in multinational firms than in SMEs (Horváth and Szabó 2019). Despite this, SMEs still have a great deal of potential to implement digital transformation more quickly. Indeed, most of the studies in the economic geography literature focus on the process of SMEs' adaptation to Industry 4.0 and cluster dynamics (e.g., Dyba, Di Maria, and Chiarvesio 2022; Götz 2020; Hervás-Oliver et al. 2019, Hervás-Oliver 2021; Isaksen, Langemyr Eriksen, and Rypestøl 2020; Isaksen et al. 2020). Nevertheless, SMEs face significant barriers and risks in the transition to Industry 4.0 and digitalization, even though they seem to have an advantage in some respects.

The main barriers to digitalization are largely due to shortcomings and incompatibilities existing in financial capabilities and support, human resources, willingness to change, lack of acceptance of the technology to be adapted, legislation, customers, level of cooperation, and strategy formulation (Horváth and Szabó 2019; Kıyak 2023). One of the most important barriers is the lack of funding, as the creation of assets to enable digitalization is a financially challenging process for firms, which is particularly true for SMEs that do not have sufficient funding for digital transformation. For example, Jankowska (2022) found that the most important internal barriers to the adoption of Industry 4.0 technologies in 16 regions at the NUTS 2 level in Poland were insufficient administrative processes and capital expenditure. A related barrier is also the lack of financial support and incentives from the government and the inability to disseminate information about government support within the cluster (Türkyılmaz et al. 2021). Another barrier is a lack of knowledge about what technology is useful and what kind of equipment and technology to invest in (Stentoft et al. 2021).

The need for a qualified labor force increases as firms adopt disruptive technologies and become more data-driven. In this case, a lack of qualified employees is one of the key barriers to Industry 4.0 (Raj et al. 2020). Therefore, there is a need to have strong human resources in terms of human capital that can adapt to disruptive technologies and have the competence to use them. Precisely for this reason, the lack of qualified employees was identified as one of the most important barriers to the adoption of 4.0 technologies (Bellandi et al. 2020; Türkyılmaz et al. 2021). One of the most important aspects for a firm to be able to adapt to technological development and change is to have a specific strategy by which firms take into account their objectives, capabilities, customer demands, and processes. Kıyak (2023), in his study of the manufacturing industry in the TR32 NUTS 2 regions of Turkey, found that 83 percent of firms do not have a strategy for digitization. Additionally, a barrier as important as any other is that of customer demand. The problem here is that firms are so focused on delivering quality products to their customers that they neglect to invest in digital technologies and experiment with new business models (Fromhold-Eisebith et al. 2021).

While there are several social and economic benefits associated with the incorporation of digitalization and disruptive technologies into the existing knowledge base, there are also some risks. One such risk is that Industry 4.0 developments may increase inequality and disrupt labor markets (Raj et al. 2020). In particular, the digitalization of the manufacturing sector creates a structural change in the labor market, increasing its mobility across sectors. At the same time, with the widespread adoption and implementation of Industry 4.0 technologies (e.g., robots and AI), there is a risk that they may have a negative impact on employees' wages (Acemoglu and Restrepo 2020). Additionally, it has been suggested that in the next quarter century, half of the current workforce, particularly those in routine jobs, may be at risk of losing their jobs due to automation and digitalization (Arntz, Gregory, and Zierahn 2017; Zemtsov 2020). Frey and Osborne (2017) estimate how many jobs could be replaced by digitalization. They find that about 47 percent of total US employment is at risk of being substituted by ICTs within the next 20 years. Although this is an advantage for companies, it increases uncertainty for the labor force. However, the positive view is that the training of employees in digital technologies will both improve product quality and reduce the cost of unskilled labor and the risk of error (Hervas-Oliver 2021). Although there is a grain of truth in both views, it is certain that the combination of Industry 4.0 technologies and skilled personnel will impact the balance of the labor market. In the context of digitalization and Industry 4.0 technologies, another area of risk is data security and firms' overdependence on technology. This is considered an important risk that prevents the use of cloud-based applications (Sommer 2015; Oztemel and Gursev 2020).

Another risk is related to the location of Industry 4.0 technologies. These technologies will initially be in regions with high R&D spending by large industrial firms, which will increase regional disparities (Jankowska et al. 2022). This puts regions at risk of exacerbating their already pronounced spatial inequalities (Balland and Boschma 2021; Fraske 2022; Jankowska et al. 2022). In this respect, it is stated that digital capabilities, which are important for the development and management of new digital technologies, are unevenly distributed among the countries and regions of the European Union, increasing the risk of polarization (Caravella et al. 2023). Some studies have observed (Balland and Boschma 2021; Jankowska et al., 2022; Laffi and Boschma 2022) that the adoption of Industry 4.0 technologies is facilitated by the fact that firms in the regions already have an Industry 3.0 technology infrastructure. Nevertheless, in the transition to Industry 4.0 technologies, most regions with weak knowledge infrastructure (e.g., in the EU region) are still far from Industry 4.0 technologies (Balland and Boschma 2021; Kiss and Nedelka 2020; Laffi and Boschma 2022). In short, Industry 4.0 and the digitalization process have the potential to exacerbate current regional disparities and spatial inequities. To mitigate this risk and minimize regional disparities, regions develop regional policies and supports that will help the transition to Industry 4.0 (Hervas-Oliver et al. 2019; Sandulli, Gimenez-Fernandez, and Rodriguez Ferradas 2021). The risks that firms see in the adoption and implementation of digitalization and Industry 4.0 also include the uncertainty of economic benefits, the quality of the workforce, high costs, investing in the wrong technology, legal aspects, and the immaturity of technological tools (De Propris and Bailey 2020; Müller, Buliga, and Voigt 2018; Sommer 2015). These risks have the potential to slow down or even prevent the transition of firms to digitalization and Industry 4.0.

9.3 Case Study and Methodology

9.3.1 Case Area: The City of Gaziantep

There are two main reasons for choosing Gaziantep as a research case: first, the city is located in one of the less developed regions of Turkey, yet many studies highlight Gaziantep as an example of success or a model of local development in Turkey's "new industrial district", the so-called "Anatolian Tigers", which industrialized rapidly in the 1980s and 1990s based on endogenous resources in some Anatolian cities (such as Denizli, Kayseri, Çorum, and Konya) outside Turkey's "traditional industrial regions" (Bayırbag 2010). Second, understanding how an industrial cluster, where informality is a common way of doing business and where labor-intensive and low-tech industrial production is concentrated, responds to the digital transformation process makes the city an interesting example. Gaziantep has a strong cluster quality in the Porterian sense, especially in the textile sector based on machine-made carpets (Oz 2004), which has become part of the global production network through its rapid industrialization since the mid-1980s and has become one of the new industrial districts where low-tech, labor-intensive industrial production is concentrated in the 2000s by organizing its unique local resources, informal institutional features, networks, and forms of entrepreneurship for local economic development within the flexible production system based on SMEs (Eraydın 2002).

Gaziantep is the largest metropolitan city in eastern and southeastern Anatolia of Turkey and the economic center of the region, with a population of 2 million. It is the seventh most important industrial city in Turkey (the largest industrial cities are, respectively, Istanbul, Ankara, Izmir, Bursa, Kocaeli, Tekirdağ, and Gaziantep) and thus the largest industrial region in eastern Turkey. Among Turkey's 81 provinces/cities, Gaziantep ranks 9th in terms of population, 8th in terms of GDP, 7th in terms of industry, and 6th in terms of exports, according to 2020 data. As of 2020, Gaziantep contributes to 2% of the national GDP and 4% of the country's manufacturing GDP. In 2020, the per capita income in Gaziantep was 6763 US dollars, while its exports totaled $8.1 billion (STB 2023).

As of 2020, Gaziantep had 3389 manufacturing enterprises operating in the industrial sector, employing a total of 131404 employees. More than half (56%) of the city's industrial production and employment is in textile manufacturing, while the rest is concentrated in traditional labor-intensive sectors such as food, plastics, leather, and paper. One of the most striking structural features of the city's industry is that 80 percent of manufacturing is low-tech. This makes Gaziantep a low-tech industrial district with a very high level of low-technology compared to the national average. In addition to being a typical industrial district, according to the 3-star cluster analysis based on 2020 data, Gaziantep has two strong textile clusters specializing in both carpet manufacturing and leather manufacturing. In this respect, the city is seen to have a mature textile industrial cluster. In fact, the fact that 54% of Gaziantep's employment and 62% of its value-added are in the textile industry is a good indicator of Gaziantep's cluster strength (TUIK 2020). As of 2019, Gaziantep, with 378 carpet firms, represents 43 percent of the carpet companies in Turkey, 76 percent of the employees in the carpet sector in Turkey (27818 employees), and 80 percent of the value-added ($630 million) in carpet production in Turkey (TUIK 2020). The most important export markets for Gaziantep are the USA (40 percent), the EU (20 percent), countries in the Middle East (20 percent), and other countries (STB 2023). Gaziantep is home to a number of key players supporting the carpet sector, including professional and lobby organizations such as the Chamber of Industry, the Exporters Association, the Model Factory, the university's textile engineering department, and training centers. Also, there are strong suppliers of synthetic yarns for machine-made carpets. Overall, the carpet sector/cluster, which is the locomotive of the Gaziantep industry, has an important position worldwide, together with its sub-industries and related and supporting institutions.

9.3.2 Data and Methods

This study is structured as a case study, one of the qualitative research designs. The study's aim is not to present explanatory research on the topic under consideration but rather to describe and understand the study topic as it is designed as exploratory research. In this primary data-based study, firms in the Gaziantep carpet industry, which were susceptible to digital transformation, were included in the sample, and interviews were conducted with the managers of firms with high digitalization potential. In general, through direct and indirect connections in a similar network, the selection of the interviewed firms was based on the recommendations of the interviewees. The research is based on the statements made by the firms operating in the carpet cluster and the representatives of the Gaziantep Chamber of Industry and other non-firm actors. During the interviews, the participants were asked, on the one hand, questions related to awareness, preparation, competence, and adaptation process toward digitalization and Industry 4.0 technologies, such as "Does your company have a vision, preparation, and strategy for digital transformation? In which areas do you use digital and Industry 4.0 technologies in your factory?" On the other hand, some other questions related to their motivations and challenges were asked, such as "What are the factors and opportunities driving you toward this process?" and "What barriers and risks do you face in the digitalization process and what are you doing about them?"

The interviews were conducted face-to-face in Gaziantep between June and November 2023. In-depth interviews were conducted with a total of 16 people, including 8 company representatives/owners, 4 officials from the Chamber of Industry and its affiliates, 2 academics from Gaziantep University, 2 experts from the Exporters Association and the Development Agency (see Table 9.1). The interviews lasted from 45 minutes to 2 hours and 12 minutes. All 8 firms that participated in the semi-structured interviews are large firms. While one of the firms in the sample is a foreign partner, the others are family owned. With the exception of one of the firms in the sample (12 years), the others consist of firms that have been operating in the organized industrial zones of the city for at least 25 years. In fact, the dates on which the firms first started operating vary between 1955 and 1998. The turnover of all these firms is between 10 and 100 million USD. Basic demographic information about the interviewees is presented in Table 9.1.

Toward the end of the field study, in line with Strauss and Corbin's (2014) explanations, the data collection process was stopped when the answers given

Table 9.1 Information about the interviewed participants.

Participants code	*Gender*	*Age*	*Work experience in the firm/ organization (year)*	*Professional experience (year)*	*Position in firm/ organization*
I1	Male	42	24	24	Firm partner
I2	Male	40	20	20	General manager
I3	Male	41	6	15	Factory manager
I4	Male	43	1	20	NGO representative
I5	Male	40	13	13	NGO representative
I6	Female	44	4	19	NGO representative
I7	Male	35	5	12	Factory manager
I8	Male	46	22	22	Factory manager
I9	Male	43	14	19	General manager
I10	Male	34	1	10	Sales manager
I11	Male	48	7	22	Firm owner
I12	Male	49	11	21	Expert
I13	Female	43	8	22	Academician
I14	Male	47	17	25	Academician
I15	Male	42	18	20	NGO representative
I16	Female	35	10	11	Expert

during the interviews became repetitive (or reached saturation). The responses given by the participants during the interviews were recorded. Transcriptions were then made of the data collected from the participants' voice recordings. This resulted in 287 pages of transcribed interviews being produced. The interviews analyzed by content analysis were first coded, then themes were identified. In the final stage of analysis, the data were interpreted by linking the data to the conceptual framework and supporting the participants' opinions with quotations.

9.4 Findings

The empirical findings of this study, which reveal the transformation process of Gaziantep carpet industry toward digitalization and Industry 4.0, are examined under two main headings. Firstly, the results on how firms define and perceive digitalization and Industry 4.0 are presented. Secondly, the barriers faced by firms in the digitalization process and the emerging risks are discussed.

9.4.1 Understanding the Concepts of Digitalization and Industry 4.0

How the interviewees define and perceive digitalization and Industry 4.0 is important for firms to be able to digitize in business and manufacturing, or to adopt and use Industry 4.0 technologies. Although the participants cannot clearly define the difference between digitalization and Industry 4.0, they define digitalization as facilitation of production and life (I7, factory manager), efficiency, minimization of errors, electronic environment and dark factory (I1, firm partner), digitalization of the entire process and products starting from human resources (I9, general manager). In addition to this definition, it can be understood that some participants use automation, which is the focus of Industry 3.0 on the basis of digitalization, and Industry 4.0 technologies on the basis of cyber-physical systems interchangeably.

> *When we say digitalization, we mean a completely electronic environment, for example, we always dream of unmanned production, which we call dark production, dark factory.*
>
> *(Interviewee, I1)*

Although the respondents do not distinguish very well between Industry 4.0 technologies and digitalization, it is clear from their statements such as "making more, trouble-free products" (Interviewee, I3), "reducing error rates in the product" (Interviewee, I10) and "increasing productivity" (Interviewee, I10) and "reducing labor costs" (Interviewee, I3) that they are aware that disruptive technologies will create benefits in the production process.

> *This is our main goal, making production more efficient and reducing the workforce.*
>
> *(Interviewee, I3)*

When analyzing the technology structure of the firms in Gaziantep Industry, it is found that Industry 3.0 technologies are widely used. According to the interviewees, using Industry 3.0 technologies such as ERP for high production capacity and maintaining this capacity in a stable way is questioned among the large firms in Gaziantep Industry. Initiatives to use some Industry 4.0 technologies such as smart sensors, cloud, cybersecurity, and data collection and analysis have also been taken by these firms.

> *[The machines] are semi-automated. That means they are labor-intensive and require human power (…). Our main goal is what I just mentioned, to increase digitalization or automation. Reducing the human element involved in production.*
>
> *(Interviewee, I2)*

Looking at whether firms invested in Industry 4.0 during the recent period, we see that they took steps toward digitalization rather than Industry 4.0.

> *We have actually taken 1–2 things for digitalization. For example, if we see that there is a deficit somewhere, we evaluate it right away. Let me put it this way: we used to buy labels from outside, now we have bought our label printing machines. As an extra second process, we have defined what they*

> *do while they are producing barcodes. There, we can document the information we need, we can address them individually.*
>
> *(Interviewee, I3)*

It is a sign of how easy it is for firms with Industry 3.0 infrastructures to embrace Industry 4.0 technology. However, it also shows that full automation has not yet been achieved in the carpet sector and that human labor is still needed, as can be seen from the following participant statement.

> *By the way, this smart sensor is not on the loom, but on the carpet preparer in what we call the garment department. It sews, bends and prepares the carpet. This machine does it all by itself. We have one of these.*
>
> *(Interviewee, I1)*

9.4.2 Barriers and Risks to Digitalization and Industry 4.0

In adopting and implementing digitalization and Industry 4.0 technologies, innovation/digitalization strategies and roadmaps at both regional and firm levels are crucial. Although manufacturing firms in Gaziantep state that digitalization is important for them, they do not have a planned and written strategy for their current and future activities and investments. Nevertheless, firms often reveal that they have unwritten strategies, but these strategies are related to the digitalization of production, rather than to investments in Industry 4.0. On the other hand, due to the economic crisis in Turkey, firms have stated that their strategies are not applicable.

> *I mean, we followed the strategy normally. We had a ruler that we were going to invest in. In other words, in the past years, we had such a ruler showing things like software, machinery, etc. We have followed this ruler. (…) We adjusted our finances accordingly. Well, since last year it's all gone. I mean, why has it gone? It's because we didn't have the foresight. The inflation, the fact that the banks are against us, that is, the total reduction of resources, the end of all incentives, the end of all investments, the fact that if you have the power, if you have the capital, you go and invest, has completely blocked our way.*
>
> *(Interviewee, I1)*

One of the most important barriers in the digital transformation process is related to the human resource issue. The lack of qualified labor to manage digital processes is also an obstacle, although industrial firms in Gaziantep are willing to go digital due to rising labor costs. However, they state that if this problem is not solved, digitalization will be meaningless for the industry.

> *(…) Obviously, digitalization is an area that makes everyone's work and life easier, but we need skilled workers and personnel who can understand it.*

> *We have some difficulties with that. (…) So if we can't find qualified personnel, digitalization is not very important for us because we can't explain it to them.*
>
> *(Interviewee, I7)*

In addition, it is understood that there is a significant problem of access to skilled labor, as trained professionals migrate to the western regions of the country or to developed countries where there are better opportunities, due to both the lack of social opportunities in the city and the inability to achieve what they want in terms of income in the industry. As a consequence of this process, one of the major obstacles to the technological transformation of firms in the Gaziantep carpet sector is the lack of competent personnel or service to solve a problem that occurs both in the use of digital machines and in the application of software programs. In this way, when it comes to repair and maintenance, the firms must bring in technical staff from the more developed regions of Turkey or from abroad.

> *We need qualified people. The quality of our human resources is very low. I'm talking about Gaziantep in general and I'm talking about the migration from the surrounding provinces, the number of people coming from the surrounding provinces are uneducated (…).*
>
> *(Interviewee, I7)*

Resistance to technological change by workers who do not want to change the way they work is another important situation. As a general idea in the manufacturing industry in Gaziantep, the existing workforce is afraid that they will be unemployed with the digital transformation. Therefore, they slow down the work or cause problems in practice. This situation is also an indication of the lack of information sharing and training.

> *We have come across some of our sectors who are trying to stop us from doing this. Either they don't need to invest, or they misapply or don't apply as much as possible when producing in the field. There are employees who behave in this way. Not on the side of the engineers, but on the side of the labors.*
>
> *(Interviewee, I6)*

Access to finance is one of the most important barriers to digitalization for firms. Significant financial resources are needed to invest in Industry 4.0 technologies. In most cases, firms that do not have sufficient equity capital are able to realize their technological transformation by taking advantage of bank loans, financial leasing, and public support. However, the most difficult financial challenge that firms face is the high interest rates that banks charge in Turkey.

We are loan borrowers, loan applicants. Everybody does it. You start working and with the profits you earn you pay all the interest on these loans, the loans themselves. So it takes 25 years, 20–25 years to pay it off. Then the conditions change and you die from the beginning, your life is spent paying off loans.

(Interviewee, I1)

Therefore, applying for incentives in order to benefit from public financial support or tax incentives is another way in which firms apply for financial resources. But here, too, government support for the region's digital transformation is low and is transferred to some firms through favoritism.

The state is not supporting us, but its own cronies. For instance, X companies, I don't know, companies that stand by the government. I mean, we are on government side too, but it is not that the state is creating incentive or not.

(Interviewee, I1)

Costs, as I said, are high, state incentives were low in these regions (...) This and that has been done, but generally our state has not done anything to reduce our production costs.

(Interviewee, I9)

Firms that have difficulties in accessing financial resources try to renew their technology by selling their obsolete machinery abroad and generating some capital for themselves, and they also want to take advantage of the investment incentives offered by the government. It is interesting to note that the investment incentives given by the government to the companies create a risk for the sector.

When I buy [new] machines, I sell them to countries with low labor power. Uzbekistan, India (...), we have realized [that] this is a risk. You are creating a competitor for yourself! (...) Let's say I've outperformed them in production for 30 years, but if a man outperforms me in 60 hours of work, I'll fall behind. The customer may prefer him. This is a big risk for companies in the long run.

(Interviewee, I9)

As a result, the carpet firms that want to benefit from the incentives for the purchase of new machines are selling the machines that they have to foreign countries. It is assumed that this situation will lead to the closure and loss of the sector by creating competitors in the carpet cluster before the country. This development is an indication of the fact that the possibility of a spatial relocation of machine carpet production is high.

The lack of support services for digital machines is another barrier to digitalization. Particularly in the follow-up processes of the purchased machines, the firms interviewed stated that they could not find support services. This situation is also an indirect barrier to the digital transformation process.

> *Of course, we can't find [supporting services] in Gaziantep. There are almost no people in Gaziantep, which can intervene in the CIP system and change the system. They come from Istanbul to help out, but we need to ask Istanbul. If they are not able to do it, then we get support from abroad.*
>
> *(Interviewee, I7)*

One of the most important elements for the firms in the development of their own business is the exchange of knowledge, which can be achieved both by getting from firms within the cluster and through access to global knowledge. In this context, the lack of co-operation between the enterprises in the carpet sector in Gaziantep and with the institutions is a situation that creates risks. Another significant reason is that there is very strong competition between the firms because they are producing the same technology and the same product.

> *There is no cooperation. Unfortunately, both in terms of production and sales, the firms are in fierce competition. The reason for this is that the carpet business is a business that requires serious investment costs in order to operate in the sector. In order to get a return on that investment, you have to sell. In order to sell, you have to be able to compete. In order to compete, you have to have power. Firms, both those that have power and those that do not, compete and influence each other to sell cheaper goods, to sell the product that you produce better. This causes serious damage to the profitability of the sector.*
>
> *(Interviewee, I9)*

Finally, the network and structure of relationships in the context in which the industry is located is another factor in the adoption of digitalization and Industry 4.0 technologies. The firms interviewed emphasize that they would move more quickly toward digitalization and Industry 4.0 if they were in a different context such as the developed western region of Turkey.

> *Maybe we could have had these studies 3–5 years ago, we could have been a step ahead. But, you know, there is a speed from west to east. The technologies used in Thrace, Marmara and the Aegean came much later to these regions. It is the same in terms of machinery, digitalization and advice. Here, you have to convince people, it doesn't matter if they are subordinates or superiors, you have to convince them.*
>
> *(Interviewee, I2)*

Therefore, due to the strong role of informal institutions such as the culture/mentality and way of doing business in the Gaziantep industry, context/space emerges as an important factor. In this respect, we see that the process of digitalization and transitioning to Industry 4.0 does not only depend on the digitalization of technological tools, but also that many factors that support and accelerate it need to be present at the same time.

9.5 Discussion and Conclusion

This study reveals the barriers and risks faced by the firms in the Gaziantep carpet industry in the process of adopting digital transformation (see Table 9.2). In doing so, the study, which has an exploratory nature, contributes to the economic geography literature on Industry 4.0 as put forward by Fraske (2022) and the spatiality of digitalization as assessed by Haefner and Sternberg (2020). One of the most important findings of the study is that firms in Gaziantep are taking steps toward digitalization rather than Industry 4.0. Although firms are primarily aware of digitalization and Industry 4.0, they cannot clearly distinguish between the two. Collective actors (Bellandi et al. 2020; Hervas-Oliver 2021) and cooperation with local institutions and universities in Gaziantep are needed to overcome this lack of knowledge and raise awareness. On the other hand, as previous studies have shown (Balland and Boschma 2021; Jankowska et al. 2022; Laffi and Boschma 2022), firms with Industry 3.0 infrastructure in Gaziantep carpet manufacturing can more easily integrate with Industry 4.0 technologies.

There are a number of *barriers* that may hinder the transition to digitalization in Gaziantep carpet industry (see Table 9.2). The main barriers to the technological transformation of local manufacturing firms in Gaziantep are lack of strategy, lack of access to finance (high loan rates, low equity, insufficient public financial support), shortcomings in government policy (favoritism due to lobbying and misdirection of incentives), lack of skilled labor, lack of support

Table 9.2 Barriers and risks to digitalization in Gaziantep carpet industry.

Category	*Barriers*	*Risks*
Human resources	Lack of qualified labor force Resistance of the labor force to technological change	Migration of skilled labor force Long learning/training time of workers
Financial resources	Lack of access to financial resources High inflation, interest, and loan rates Insufficient public financial support	Unequal distribution of state incentives through nepotism/favoritism Low equity capital of firms
Contextual and/or Institutional factors	Lack of (digital/innovation) strategy Lack of technological knowledge and sharing Lack of support services Lack of clustering platform	Lack of trust and cooperation Limitations of the context/city through cognitive inertia and institutional isomorphism Shifting of the market to different countries Lack of governance

services, lack of cooperation, and limiting effect of context. On the other hand, firms and regions face serious uncertainties and *risks* due to the rapid pace of technological change. Written strategies and digital roadmaps are a crucial factor in adopting and implementing digitalization and Industry 4.0 technologies (Balland and Boschma 2021; Fraske 2022; Jankowska et al. 2022; Raj et al. 2020). When this factor is considered in the context of Gaziantep industry, firms' strategies are mostly short-term and situational, and their production and investment decisions are not based on a long-term, planned strategy, which is an important obstacle. One of the most important barriers for the firms in the digitalization process is the access to financial resources (Horváth and Szabó 2019; Kıyak, 2023; Türkyılmaz et al. 2021). As investments in digital technologies require high capital, firms need financial resources other than their own equity. However, due to inflation, firms in Gaziantep's industrial sector are not able to benefit from both government incentives and bank loans. Another situation is the unequal distribution of government incentives among firms due to favoritism. This situation hinders and demotivates firms that want to invest in digitalization.

On the other hand, government incentives are an opportunity for firms. However, they run the risk of losing the market, especially because firms that want to benefit from investment incentives sell their old machines abroad. One of the most important factors that has emerged in the process of digitization and the transition to Industry 4.0 in the case of Gaziantep is the lack of a qualified labor force (see Table 9.2). This situation is not unique for Gaziantep, but is also an issue in other countries and regions, including Turkey (Kıyak 2023), the Larciano region of Italy (Bellandi et al. 2020), Poland (Jankowska et al. 2022), and Kazakhstan (Türkyılmaz et al. 2021). A risk to technological transformation is the migration of a relatively small number of skilled workers to places with better opportunities. It is understood that a culture built on space/context affects innovation and production processes in the transition to digitalization and Industry 4.0 technologies. In this regard, the lack of trust and cooperation among firms in Gaziantep's industry and the lack of governance among local actors are among the main risks to the realization of technological transformation. On the other hand, since each place or region has its own unique structure, it is understood that firms in Gaziantep can realize technological transformation more easily and faster, and thus be more competitive, if they are located in the developed western regions of the country. Similarly, the study by Fromhold-Eisebith et al. (2021) for Germany highlights the importance of context. While the fast pace of technology facilitates national and global cooperation opportunities for manufacturing firms and the city of Gaziantep, strong local cooperation also needs to be developed to compete in the global market. While trust-based collaboration networks play an important role in firms' adoption of digitalization and Industry 4.0 technologies, as seen in case studies from the EU (Balland and Boschma 2021) and Germany (Götz 2019), this is not the case in Gaziantep. This finding is quite remarkable in the cluster and Industry 4.0 literature.

To this end, the barriers and risks faced by firms could be reduced by developing a common platform and technological transformation strategy among industrial actors in Gaziantep through a place-specific governance/cluster organization (Bellandi et al. 2020; Hervas-Oliver et al. 2019; Jankowska et al. 2022; Micek et al. 2022; Sandulli, Gimenez-Fernandez, and Rodriguez Ferradas 2021). However, in the absence of such collective initiatives, some sectors in Gaziantep industry may face the risk of lock-in, as in Germany and Korea (Hassink 2010) and the Prato textile cluster in Italy (Bellandi et al. 2020). It is important to note that in such a situation, or if Gaziantep cannot keep up with technological change, different industries in Gaziantep may face the risk of being left out of the global value chain (Völlers et al. 2023). On the other hand, we see that product diversification has been initiated in various sectors to counter the risk of Gaziantep's industry being locked in, and a similar situation has been experienced in the Viareggio yacht building cluster in the Tuscany region of Italy (Bellandi et al. 2020). One of the most important sources of digitalization is sharing global and local knowledge (Hervas-Oliver et al. 2019; Jankowska et al. 2022; Winter 2020). However, especially in Gaziantep industry, the lack of cooperation and trust among firms and between firms and organizations means that knowledge sharing remains limited. The general culture and informal institutions of the city, the mentality and mindset of the industrialist, the way of doing business, and its networked structure, in other words, the context/space itself, pose a significant risk to the technological transformation of Gaziantep industry.

In summary, firms in Gaziantep industry are both aware of digitalization and Industry 4.0 technologies, and those who adopt and use these technologies are aware of the benefits, barriers, and risks of technological transformation, as stated by Jankowska et al. (2022) in their study. It is understood that there has been an acceleration of technological transformation in Gaziantep industry, especially in relation to customers, and at the same time, there have been significant changes in productivity, quality, efficiency, and competitive position (Lazzeretti et al. 2023; Winter 2020). On the other hand, as in the study of Hervas-Oliver et al. (2019), we can see from the observed results that important professional organizations such as the Gaziantep Chamber of Industry and pioneer firms in different sectors play a key role (through traveling to and from international fairs, model factory and various digital transformation trainings) in the digital transformation journey of Gaziantep, taking on an important mission in the transformation of other firms at the local level and triggering the transformation of these firms. Therefore, this situation shows us that firms and other actors with sufficient knowledge, technological infrastructure, and expertise in digital transformation have the potential to contribute significantly.

However, as with the national scale, there are uncertainties about how the digital transformation platform will work in Gaziantep's industry. In order to overcome these uncertainties, Erdil (2023) has proposed various policy measures for Turkey. In this regard, further increasing the level of awareness with the support of the public sector, establishing a common governance platform, increasing qualified human capital with a revised education system, identifying priority areas and developing competencies, further strengthening the technical infrastructure, improving the capacities of digital technology suppliers and increasing the access of their products and services to customers, and finally developing cooperation with all stakeholders are the priority policy recommendations to eliminate uncertainty and risks in technological change (Erdil 2023). It should be noted, however, that these recommendations can be implemented locally if the conditions in Gaziantep are taken into account. The results of this study can help policy makers, decision-makers, and practitioners at both national and local levels to address the highlighted barriers and thus pave the way for the successful implementation of digitalization and Industry 4.0 in Gaziantep's manufacturing sector, specifically in the carpet industry.

References

Acemoglu, D., and P. Restrepo. 2020. Robots and jobs: Evidence from US labor markets. *Journal of Political Economy* 128:2188–244. doi:10.1086/705716

Arntz, M., T. Gregory, and U. Zierahn. 2017. Revisiting the risk of automation. *Economics Letters* 159:157–60. doi: 10.1016/j.econlet.2017.07.001

Balland, P. A., and R. Boschma. 2021. Mapping the potentials of regions in Europe to contribute to new knowledge production in Industry 4.0 technologies. *Regional Studies* 55 (10–11):1652–66. doi: 10.1080/00343404.2021.1900557

Bayırbag, M. K. 2010. Local entrepreneurialism and state rescaling in Turkey. *Urban Studies* 47 (2):363–85. doi: 10.1177/0042098009349022

Bellandi, M., E. Santini, C. Vecciolini, and L. De Propris. 2020. Industry 4.0: Transforming local productive systems in the Tuscany region. In *Industry 4.0 and Regional Transformations*, ed. L. De Propris, and D. Bailey, 84–96. Routledge.

Capello, R., and C. Lenzi. 2021. *The regional economics of technological transformations: Industry 4.0 and servitisation in European regions*. Routledge.

Caravella, S., V. Cirillo, F. Crespi, D. Guarascio, and M. Menghini. 2023. The diffusion of digital skills across EU regions: structural drivers and polarisation dynamics. *Regional Studies, Regional Science* 10 (1):820–44. doi: 10.1080/21681376.2023.2265498

Corò, G., and M. Volpe. 2020. Driving factors in the adoption of Industry 4.0 technologies: An investigation of SMEs. In *Industry 4.0 and Regional Transformations*, ed. L. De Propris, and D. Bailey, 112–32. Routledge.

De Propris, L., and D. Bailey. 2020. Disruptive Industry 4.0+ Key concepts. In *Industry 4.0 and Regional Transformations*, ed. L. De Propris, and D. Bailey, 1–23. Routledge.

Dyba, W., E. Di Maria, and M. Chiarvesio. 2022. Actions fostering the adoption of Industry 4.0 technologies in manufacturing companies in European regions. *Journal of Regional Research* 53:27–46. doi: 10.38191/iirr-jorr.22.009

Eraydın, A. 2002. *Yeni sanayi odakları: Yerel kalkınmanın yeniden kavramsallaştırılması* [New industrial districts: Reconceptualising local development]. ODTÜ.

Erdil, E. 2023. *Dijital dönüşüm ve Türkiye*. İstanbul Politik Araştırmalar Enstitüsü [Digital transformation and Turkey. Istanbul Political Research Institute], 9 September 2023, https://istanpol.org/Uploads/ContentManagementFile/dijitaldonusum_pr_240423.pdf

Fraske, T. 2022. Industry 4.0 and its geographies: A systematic literature review and the identification of new research avenues. *Digital Geography and Society* 3:100031. doi: 10.1016/j.diggeo.2022.100031

Frey, C. B., and M. A. Osborne. 2017. The future of employment: How susceptible are jobs to computerisation?. *Technological forecasting and social change* 114:254–280. doi: 10.1016/j.techfore.2016.08.019

Fromhold-Eisebith, M., P. Marschall, R. Peters, and P. Thomas. 2021. Torn between digitized future and context dependent past–How implementing 'Industry 4.0'production technologies could transform the German textile industry. *Technological Forecasting and Social Change* 166:120620. doi: 10.1016/j.techfore.2021.120620

Götz, M. 2019. The industry 4.0 induced agility and new skills in clusters. *Foresight and STI Governance* 13 (2):72–83. doi:10.17323/2500-2597.2019.2.72.83

Götz, M. 2020. Cluster role in industry 4.0–a pilot study from Germany. *Competitiveness Review: An International Business Journal* 31 (1):54–82. doi: 10.1108/CR-10-2019-0091

Götz, M., and B. Jankowska. 2017. Clusters and industry 4.0 – do they fit together?, *European Planning Studies* 25 (9):1633–53. doi: 10.1080/09654313.2017.1327037

Haefner, L., and R. Sternberg. 2020. Spatial implications of digitization: State of the field and research agenda. *Geography Compass* 14 (12):1–16. doi: 10.1111/gec3.12544

Hassink, R. 2010. Locked in decline? On the role of regional lock-ins in old industrial areas. In *The Handbook of Evolutionary Economic Geography*, ed. R. A. Boschma, and R. L. Martin, 450–68. Edward Elgar.

Hervás-Oliver, J. L. 2021. Industry 4.0 in industrial districts - regional innovation policy for the Toy Valley district in Spain. *Regional Studies* 55 (10–11):1–12. doi: 10.1080/00343404.2021.1939861

Hervas-Oliver, J. L., S. Estelles-Miguel, G. Mallol-Gasch, and J. Boix-Palomero. 2019. A place-based policy for promoting industry 4.0: The case of the Castellon ceramic tile district. *European Planning Studies* 27 (9):1838–56. doi: 10.1080/09654313.2019.1642855

Horváth, D., and Szabó, R. Z. 2019. Driving forces and barriers of industry 4.0: Do multinational and small and medium-sized companies have equal opportunities? *Technological Forecasting & Social Change* 146 (September):119–32. doi: 10.1016/j.techfore.2019.05.021

Isaksen, A., E. Langemyr Eriksen, and J. O. Rypestøl. 2020. Regional industrial restructuring: Asset modification and alignment for digitalization. *Growth and Change* 51:1454–70. doi: 10.1111/grow.12444

Isaksen, A., M. Trippl, N. Kyllingstad, and J. O. Rypestøl. 2020. Digital transformation of regional industries through asset modification. *Competitiveness Review* 31:130–44. doi:10.1108/CR-12019-0140

Jankowska, B., E. Mińska-Struzik, M. Bartosik-Purgat, M. Götz, and I. Olejnik. 2022. Industry 4.0 technologies adoption: Barriers and their impact on Polish companies' innovation performance. *European Planning Studies* 31 (5):1029–49. doi: 10.1080/09654313.2022.2068347

Kıyak, B. 2023. İmalat sanayinde dijitalleşme eğilimleri: TR32 düzey 2 bölgesi örneği. *Bölgesel Kalkınma Dergisi* [Digitalization trends in the manufacturing sector: case of TR32 NUTS II Region] 1 (02):24–38.

Kinkel, S. 2020. Industry 4.0 and reshoring. In *Industry 4.0 and Regional Transformations*, ed. L. De Propris, and D. Bailey, 195–213. Routledge.

Kiss, É., and E. Nedelka. 2020. Geographical approach of industry 4.0 based on information and communication technologies at Hungarian enterprises in connection with industrial space. *Hungarian Geographical Bulletin* 69 (2):99–117. doi: 10.15201/hungeobull.69.2.2

Laffi, M., and R. Boschma. 2022. Does a local knowledge base in Industry 3.0 foster diversification in Industry 4.0 technologies? Evidence from European regions. *Papers in Regional Science* 101 (1):5–35. doi: 10.1111/pirs.12643

Lazzeretti, L., R. B. Domenech, J. L. Hervas-Oliver, and N. Innocenti. 2023. Artificial intelligence, big data, algorithms and industry 4.0 in firms and clusters. *European Planning Studies* 31 (7):1–7. doi: 10.1080/09654313.2023.2220490

Micek, G., K. Gwosdz, A. Kocaj, A. Sobala-Gwosdz, and A. Świgost-Kapocsi. 2022. The role of critical conjunctures in regional path creation: A study of industry 4.0 in the Silesia region. *Regional Studies, Regional Science* 9 (1):23–44. doi: 10.1080/21681376.2021.2017337

Müller, J. M., O. Buliga, and K. I. Voigt. 2018. Fortune favors the prepared: How SMEs approach business model innovations in Industry 4.0. *Technological Forecasting and Social Change* 132: 2–17. doi: 10.1016/j.techfore.2017.12.019

OECD. 2018. Policy brief on the future of work: Putting faces to the jobs at risk automation. Accessed September 9, 2023. https://www.oecd.org/employment/Automation-policy-brief-2018.pdf

Oztemel, E., and S. Gursev. 2020. Literature review of Industry 4.0 and related technologies. *Journal of Intelligent Manufacturing* 31 (1):127–82. doi: 10.1007/s10845-018-1433-8

Oz, Ö. 2004. The carpet cluster in Gaziantep. In *Clusters and Competitive Advantage: Turkish Experience*, ed. Ö. Oz, 114–34. Palgrave Macmillan. doi: 10.1057/9780230512467_6

Raj, A., G. Dwivedi, A. Sharma, A. B. L. de Sousa Jabbour, and S. Rajak. 2020. Barriers to the adoption of industry 4.0 technologies in the manufacturing sector: An inter-country comparative perspective. *International Journal of Production Economics* 224:107546. doi: 10.1016/j.ijpe.2019.107546

Sandulli, F. D., E. M. Gimenez-Fernandez, and M. I. Rodriguez Ferradas. 2021. The transition of regional innovation systems to Industry 4.0: the case of Basque Country and Catalonia. *European Planning Studies* 29 (9):1622–36. doi: 10.1080/09654313.2021.1963049

Schwab, K. 2016. *The fourth industrial revolution*. Currency.

Sommer, L. 2015. Industrial revolution-industry 4.0: Are German manufacturing SMEs the first victims of this revolution?. *Journal of Industrial Engineering and Management* 8 (5):1512–32. doi: 10.3926/jiem.1470

STB. 2023. *Gaziantep İl Sanayi Durum Raporu 2021*. Sanayi ve Teknoloji Bakanlığı Gaziantep Sanayi ve Teknoloji İl Müdürlüğü [Gaziantep Provincial Industry Status Report 2021. Ministry of Industry and Technology Gaziantep Directorate of Industry and Technology] 9 September 2023. https://www.sanayi.gov.tr/plan-program-raporlar-ve-yayinlar/81-il-sanayi-durum-raporlari/mu1602011633

Stentoft, J., K. Adsbøll Wickstrøm, K. Philipsen, and A. Haug. 2021. Drivers and barriers for industry 4.0 readiness and practice: Empirical evidence from small and medium-sized manufacturers. *Production Planning & Control* 32 (10):811–28. doi:10.1080/09537287.2020.1768318

Strauss, A., and J. Corbin. 2014. *Basics of qualitative research*. 4th ed. Sage.

Szalavetz, A. 2019. Industry 4.0 and capability development in manufacturing subsidiaries. *Technological Forecasting and Social Change* 145: 384–95. doi: 10.1016/j.techfore.2018.06.027

TUIK 2020. *Yıllık Sanayi ve Hizmet İstatistikleri 2020* [TSI 2020. Annual Industry and Services Statistics 2020]. TUIK.

Türkyılmaz, A., D. Dikhanbayeva, Z. Suleiman, S. Shaikholla, and E. Shehab. 2021. Industry 4.0: challenges and opportunities for Kazakhstan SMEs. *Procedia CIRP* 96:213–18. doi: 10.1016/j.procir.2021.01.077

TTGV 2018. *Sanayide dijital dönüşüm-Eğitim* [TDFT 2018. Digital transformation in industry - Education]. TTGV.

UNIDO. 2020. *Industrial development report 2020-industralizing in the digital age*. UNIDO.

Völlers, P., N. Yavan, and M. Franz. 2021. The reaction of foreign firms to institutional changes: The case of German direct investment in Turkey. *Applied Geography* 134:102503. doi: 10.1016/j.apgeog.2021.102503

Völlers, P., T. Neise, P. Verfürth, M. Franz, F. Bücken, und K. P. Schumacher. 2023. Revisiting risk in the global production network approach 2.0-Towards a performative risk narrative perspective. *Environment and Planning A: Economy and Space* 55 (8):1838–58. doi: 10.1177/0308518X231169288

Yavan, N., Ş. Yılmaz, and A. Aniç. 2022. Institutional context and territorial policy: Analyzing the new regional policy and regional development agencies in Turkey. In *Economies, institutions and territories dissecting nexuses in a changing world*, ed. L. Storti, G. Urso, and N. Reid, 135–64. Routledge.

Winter, J. 2020. The evolutionary and disruptive potential of Industrie 4.0. *Hungarian Geographical Bulletin* 69 (2):83–97. doi: 10.15201/hungeobull.69.2.1

Zemtsov, S. 2020. New technologies, potential unemployment and 'nescience economy' during and after the 2020 economic crisis. *Regional Science Policy & Practice* 12 (4):723–43. doi: 10.1111/rsp3.12286

10 Risk Perception of Foreign Businesses in the Face of Local Contextual Conditions

Piotr Pachura

10.1 Introduction

The choice of investment location for businesses has a global dimension, but also a local dimension, one of the very important phenomena is the use of localness or the exploration of local resources and local embeddedness (Yip 1992; Yavan 2010). Business investment in new geographic areas is one of the most important forms of capital transfer. Decisions about new business locations involve various types of risk. The basic groups of risk are primarily recognized in economic and legal terms. These basic risks are related to changes in currency exchange rates, interest rates, fluctuations in the foreign market, and unpredictable changes in legal regulations. Instability in political systems and changes in a country's fiscal systems are often equally important risks. At the same time, a substantial risk factor is cultural distance, which means differences in cultural, social, symbolic capital, and the structure of social behavior (Leavy 1984; López-Duarte, Vidal-Suárez, and González-Diaz 2015; Shenkar 2012). One of the elements often neglected in risk assessment studies is the distinctive religious context of a given space in a place (Hong et al. 2023; Tocar 2019). The aim of the chapter is to present the results of research on the risk perception of a specific local sociocultural context in a location and the performance of multinational corporations.

The specificity of the research is to recognize the potential impact of a very powerful contextual religious space of place on international business locations and activities. To achieve the objectives of the study, the case of one of Europe's largest religious pilgrimage urban-hub, Czestochowa (Poland), was analyzed.

The structure of the chapter consists of a review of the literature, a description of the studied case of a city with a strong religious context, a characterization of the research methodology, a presentation of the research results, and the final conclusions. The presented study is one of the few examples of the analysis of the perception of the risk of investment of multinational corporations in the context of the determinants of the religious culture of the location space.

DOI: 10.4324/9781003457626-13

10.2 Literature Review

The theory of the location of businesses in geographic space has always had great importance in economic geography. The first theories of business location were associated with the development of the German school of economic geography (see Lösch 1940; Weber 1929). The development of later theories tended toward more holistic approaches such as the work of Isard (1956). With regard to more contemporary research, it is worth mentioning the achievements of Krugman (1991) in the field of business location theory in the context of the globalization phenomenon.

Theories of localization in terms of the importance of cultural or symbolic spatial factors also have a long tradition. Hall's research on space (1959) expanded into an anthropology of space analyzing the spatial determinants of social relations and behavior, elaborating the influential and widely discussed concept of high- and low-context cultures (1966).

Perhaps Hall's concepts could already be considered outdated due to the globalization of culture or the globalization of management, understood as the diffusion of management theory and practice through, among other things, the expansion of corporations and the mobility of employees, especially managers (Suwala, Pachura, and Schlunze 2022). For the most part, however, theories about the importance of contextual space in the location of companies are not challenged (Kittler, Rygl, and Mackinnon 2011).

A very well-known and widely discussed empirical research on the issue of context was conducted by Hofstede, who analyzed the determinants of the cultural contexts of enterprises in dozens of countries (1984, 2011). The theory of business location in social space or more broadly in cultural space is an issue that has been developed in recent decades in correlation with the expansion of transnational corporations and global supply chains and production networks (Gereffi, Humphrey, and Sturgeon 2005; Gereffi and Korzeniewicz 1994). There is a research trend in the scientific literature that links the issues of risk, foreign direct investment (FDI), and the socio-cultural dimensions of religion. As Rodríguez notes, "In economics, risks are related to losses and risk assessment is related to loss measuring. In country risk, the elements than can provoke a loss are of three kinds: economic, political and social" (2016, 2). In this cognitive perspective, Mansour and Jlassi (2014) focus on the relationship between religion and investments, financing decisions, risk perception, ethical values, and psychological behavior. Kazunobu, Kimura, and Lee (2013) empirically study the impact of various risk components on inward FDI using the example of Myanmar. Research conducted by Russell et al. (2023) analyzes the relationship of religion-based ethics, particularly Judeo-Christian ethics, in a given country to economic freedom, business activity, gender equality, social progress, and corruption. The research results show that Judeo-Christian ethics have a positive relationship with factors related to a society's sustainable business culture – i.e. greater economic freedom and activity, social progress, and less corruption. Hong et al. (2023) examine the impact of religion on FDI using empirical data from 1985 to 2019 and calculating the religious distance between the country of origin and the host country. The results indicate that FDI flows are smaller for country pairs with greater religious distance. Moreover, the negative impact of religious differences is less pronounced when there is greater religious diversity in the host country. Studies by Hong et al. (2023) suggest that both religious differences and the level of religiosity play an important role in explaining the risk of international FDI flows.

Research conducted over a decade in dozens of countries by Lucke and Eichler (2016) suggests that foreign investors prefer to invest in countries with less heterogeneous societies than their own. Research by Lee, Lee, and Oh (2023) establishes a relationship between the degree of religiosity in different U.S. regions and the use of derivatives in the hedge fund industry. The results indicate that the level of religiosity in a community is negatively related to the likelihood of hedge funds using derivatives for speculative purposes. At the same time, funds in regions with higher ratios of Catholics to Protestants are more likely to engage in speculative derivatives trading. Research on religious proximity and FDI location choice conducted by Soussane, Mansouri, and Mansouri (2022) using Moroccan companies indicates that religion is important for inward and outward FDI. They also point out that "religious similarity promotes friendly and trusting relationships, facilitates large contracts, and establishes several economic partnerships" (2022, 427). Abbas, Abbas, and Abbas (2022) studied the impact of religious freedom and FDI on economic growth in Pakistan, India, Bangladesh, and Sri Lanka. The empirical results showed that an increase in religious freedom increases economic activity and stimulates growth. The results confirm that economies characterized by religious freedom attract more FDI. Interesting research on the impact of religious freedom on risk assessment is presented by Alon and Spitzer (2003). The authors prove that religion (in this case, religious freedom) as an element of the social and cultural environment can influence the international business environment. In addition, they conclude that religion can affect the perception of a country's risk and thus influence investment decision strategies. "Religion shapes culture: the way people behave, their view on life, their history, their conceptions of right [...]" (Alon and Spitzer 2003, 54). As Alon and Spitzer further note "more research is needed on the basic concept of religion as an element of the international business environment that affects the perceptions, conduct, and profitability of foreign investment" (2003, 61).

The location of an organization in a certain space, especially in the physical space is one of the most important aspects of its existence. At the same time, the local space influences the organization, and the organization influences a surrounding space (Pachura 2021). Everything takes place simultaneously in many dimensions: economic, social or environmental, and cultural. At the same time, corporations gradually learn a context that exists in a given place and influence the structure of local companies and societies. The term "context" is used to describe the nexus of interrelated, interdependent factors that are needed for effective communication and a shared understanding of the nature of things (Pachura 2015; Suwala 2014, 2021). In studying the dynamics of economic spaces an increasingly vital role is attributed to cultural factors, "places are about context…" (Basco and Suwala 2020, 188). It seems worth mentioning such classic exemplary works as: Hall (1959, 1966), Perlmutter (1969), Perlmutter and Heenan (1974), Hofstede (1984, 2011), Schlunze et al. (2012), and Fuchs et al. (2017). Interesting research is created in the space of interfaces or in spaces that interpenetrate, Fuchs et al. (2017) emphasize the great importance of culture, especially the culture of place in location processes involving cross-border areas. At the same time, they define the specificity of these spaces as "interspace" which means "the sociocultural patterns that underlie human agency, as well as the power of rules that influence an individual's sense of identity and otherness" (2017, 4).

10.3 Częstochowa's Dual Role of a Pilgrimage and Industrial Center

Research on the importance of religious context for the location decisions and operations of foreign business was conducted using the example of one of the largest and most recognizable religious pilgrimage sites in Europe. The study area was a city located in the central-southern part of Poland – Częstochowa. The reason for choosing such a study area was the simultaneous occurrence of two phenomena – on the one hand, a unique religious capital, and on the other hand, industrial traditions and a relatively large amount of FDI. The city, as a major center of the world religious cult, is defined by the Catholic Sanctuary of the Black Madonna, which is a place of religious devotion visited by several million pilgrims a year and called "the spiritual capital of the country". The sanctuary in Częstochowa (Jasna Gora) functions as an ambiguous and multi-layered religious-mythological construction, important mainly for the history of Poland, but also placed in the group of the world's greatest shrines, such as Jerusalem, Vatican, Lourdes, and Fatima. The center of world worship is the Image of Our Lady of Czestochowa as a certain key that opens the door to the layers of national mythology. The image is treated as a sign symbol of national identity. The second interesting motif is that of the defense of the holy fortress city. It arises from the topos of the defense of Jasna Gora during the Swedish invasion and is part of the motif of defense of the holy city-temple functioning in the European culture (Niedźwiedź 2000).

Table 10.1 presents the characteristics of the number of pilgrims in relation to the number of inhabitants in the most important destinations of the Marian religious cult in Europe.

The compilation in Table 10.1 highlights the typical phenomenon that European religious centers are mostly small towns, with the exception of the city that is subject of this study. These small urban centers (such as Fatima, Lourdes, Santiago de Compostela) are basically oriented to serve pilgrimage flows and shrines are the main (if not the only) citygenic factor. In the case of these cities, it is also difficult to discuss of classic industrial business or business services investments except possibly in facilities for accommodation or

Table 10.1 Number of visitors to selected sanctuaries in Europe in 2019 and their populations in 2019/2021.

Europe's largest pilgrimage cities (towns)	*City population*	*Approximate number of visiting pilgrims*
Fatima (Portugal)	13 212 (2021)	6 300000
Lourdes (France)	13 132 (2019)	5 000000
Częstochowa (Poland)	*214 342 (2021)*	4 400000
Santiago de Compostela (Spain)	96 260 (2019)	347578

Source: author's compilation based on: Mróz 2021 and La Croix International 2020.

devotional trade. In contrast, in the case of Częstochowa city studied in the research reported in this chapter, we are faced with a completely different circumstance involving the occurrence of multiple simultaneous citygenic processes, and there are multiple city potentials. The modern shape of the city as an arena and destination for foreign business investment has been formed by its socio-economic development since the industrialization era.

The development of the city from the 19th century onward was associated with the construction of the Warsaw-Vienna railroad passing through Częstochowa in 1846 (Pawlicki 1897). The first factories were established in the city in the 1870 associated with the arrival of foreign capital, mainly French, Belgian, and German located mainly in the textile industry. At the same time, the mining and metallurgical industry was growing related to the extraction of iron ore. In the 1930s of the XX century and after World War II, the industrial development in the Częstochowa area was significant on a national scale. During the socialist era, political decisions about the city's industrial development were related to supporting the dominance of the socialist working class in large industrial plants and thus eliminating the city's religious context (Sobalski and Nowak 2017).

During the process of economic transformation since the 1990s, the city has seen the collapse of numerous industrial plants, effect of the processes linked to the economic reconversion of the subregion is the appearance of unemployment and brownfields. The basic data for 2020, are as follows: the level of the city's own investments is 33 million euros, the unemployment rate in the city is 4.2%, 6 million euros invested in business infrastructure, direct business investment (25 new enterprises) amounted to 273 million euros (City of Częstochowa Official Website 2023), city's international rating (Fitch Ratings) is BBB+. Thus, the analyzed case of a city with a strong (relatively) local economy and at the same time a very explicit religious context seems appropriate to try to address interesting research questions.

10.4 Materials and Methods

At the beginning of the methodology development, one can consider: *What do we intuitively suspect?* In the case of small cities (religious centers), there is no connection between the sacred function and FDI because there is too little endogenous potential, or this relationship is evident because FDI is directly related to this religious function of the place. In the case of a larger city with greater endogenous potential and many city-forming functions, such a relationship is more difficult to identify.

To go beyond the intuitive assumption, the following research question was posed: *Is it possible that the endogenous urban strong religious context of the place – can be a risk factor in the perception of foreign businesses investing and operating in this location?* Risk here is understood primarily as religious and cultural distance and the religious city-forming function of the place. Confirmation or falsification of this question can contribute to the state of knowledge on the geography of business location and the impact of contextual space on the performance of foreign investing businesses.

The tool for collecting primary data was in-depth interviews conducted in person and via online videoconferences (from 04.2022 to 11.2022). The research was carried out in a group of the following companies, located within the administrative borders of the city (see Table 10.2).

In this research, within the framework of the in-depth interview technique, a semi-structured interview was used. This type of interview requires so-called dispositions for the interview, i.e. a list of the researcher's information needs, but these can be formulated relatively freely and in general terms (Bryman 2012; Jupp 2004). In this case, the interview scenario was structured based on the distinction of three levels of decisions and performance of the organization, i.e. strategic, tactical, and operational.

The surveyed companies (see Table 10.2) had two types of locations, one type being locations within the city in conjunction with standard city infrastructure, and the other type being company locations in special economic investment areas (SEZs) also located within the city. At the beginning of the socio-economic transformation of the countries of Central and Eastern Europe, including Poland, the idea of SEZs as an instrument of economic policy gained great importance. SEZs are separate areas within the territory of

Table 10.2 List of surveyed enterprises (author).

Enterprise	*Country of origin*
ZF Friedrichshafen AG	Germany
Stoelzle Glass Group	Austria
Comptoir General du Ressort S.A	France
Hillwood Investment Properties	United States
TS Tech Co., Ltd	Japan
Liberty Steel Group	United Kingdom
Cooper Standard Automotive Inc.	United States
TRW Automotive Holdings Corp	United States
Brembo S.p.A	Italy
Guardian Glass LLC	United States

Poland, where businesses can operate under more favorable conditions than in the rest of the country. Until 2018, the zones operated as separate territories, as of 2018, the entire territory of Poland is covered by regulations that allow to obtain investment incentives similar to those in SEZs. This solution to support the attraction of FDI is called the Polish Investment Zone and it is an instrument through which it is possible to obtain tax benefits for the implementation of a new investment. A foreign company can use it throughout Poland, regardless of the size of the company. Support is granted to companies operating in the industrial and modern services sectors (Act of Parliament, 2018). A characteristic feature of SEZs is territoriality, i.e. limiting the availability of investment benefits to specific areas. In the case studied, the city of Częstochowa has two SEZs with a total area of 70 hectares, where 33 enterprises operate, 12 of which are foreign capital investments (City of Częstochowa Official Website 2023). On the territory of Czestochowa, there is a part of the Katowice Special Economic Zone, which has been repeatedly recognized as the best SEZ in Europe and the second in the world (FDI Business Financial Times 2019).

The research was designed in three areas or otherwise at three organizational levels. The most classical approach to organizational structure and management derived from the processes of standardization and job specialization is based on the existence of three levels (Mintzberg 1979). This is a management hierarchy related to the responsibility and complexity of decisions and their level of risk at three levels: top-level (strategic), middle-level (tactical), and lower-level (operational). Top-level managers are responsible for the strategic decisions and risks relating to the setting of organizational goals and the implementation of the main objectives of the organization. Middle-level managers are responsible for achieving the set strategic goals, and lower-level managers are responsible for the operations of each business unit (Drucker 1954; Hodge, Anthony, and Gales 1996; Jackson and Morgan 1982) (Figure 10.1).

In addition, five in-depth interviews were conducted with managers of public service agencies that support foreign investment in the city (municipal authority, regional development agency, investor service office, special economic zones). At this stage, the research was performed in the most important centers related to the economic development, support and stimulation of

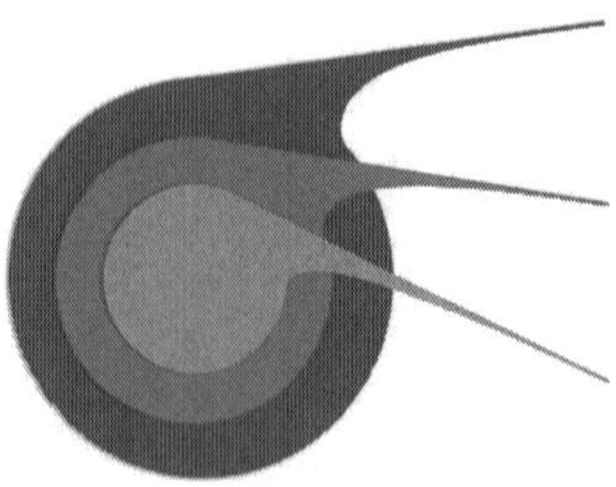

Figure 10.1 Three levels of results analysis of corporate managers' surveys (author).

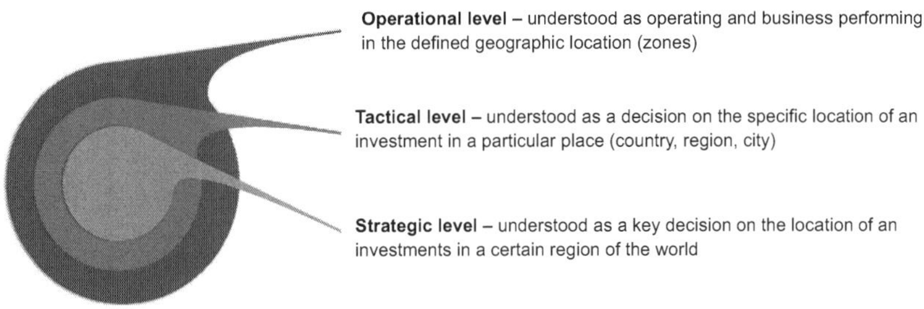

Figure 10.2 Three levels of results analysis of city administration and agencies surveys (author).

entrepreneurship, attraction of economic investors and partners, support of start-ups, and creation of infrastructure for FDI (Figure 10.2).

The analysis of the primary data from the interviews was based on a transcription of the interviews and then thematic content analysis. Then a screening of the content was done in order to identify some occurring patterns in the data set. At the end of the procedure, the cognitive categories corresponding to three levels of complexity (organizational performance) were adopted: macro (strategic), meso (tactical), and micro (operational), which allowed a structured presentation of the results.

10.5 Results of Empirical Research

During the interviews, managers of international companies indicated that at the level of strategic localization decisions, the risks related to the economic stability of a given location are taken into account first and foremost. The factors taken into account are mainly those related to a specific world region, in this case, Central and Eastern Europe, which are also part of the European Union economic zone. The availability of a qualified group of human resources, labor costs, and the size of the regional sales market are also important risk factors at the level of strategic decisions. *"Such decisions are primarily an analysis of conditions more on a macro scale – market, infrastructure, human resources. Religious or cultural issues not necessarily" (interview with US company, business development manager, 2022)*. The studies indicate that at the strategic level, the contextual risk factor in this case related to the religious context has no significance. Managers state that there is always some local context and that it is not significant as a strategic risk factor for investment location, although there is often an awareness of the presence of this context, especially among European companies. *"We make strategic decisions, rather directional we don't get too much into some very local issues" (interview with UK company, production development manager, 2022)*. Especially since the localization takes place in the broader sphere of European culture and the conditions of the European Union.

The interviews show that at the tactical level – understood as a decision on the specific location of an investment in a particular place (country region, city) – the religious context has significance as the existence of a local brand of the place, which in a way causes geographic and socio-cultural identification of the location. It turned out to be more important for managers from non-European countries (Japan, United States). It seems that a religious branding can mitigate investment risks. Moreover, beyond the strategic level, it turned out that, awareness of the culture and symbolism of a place positively influences the brand perception of a specific location. Most often, such a cultural brand of a place is associated with the historical heritage (such as Krakow), the heritage of industrial culture (such as Katowice), but also spiritual culture, as in the studied case of Częstochowa. To some extent, this reduces the perception of investment risk. *"Yes, if a place, a location has a good connotation, stands out with that connotation it's rather good for investment" (interview with French company, human resources manager, 2023)*.

At the operational level – understood as business operations and activities in a defined geographic location (zones, city) – the religious context of the place is partly important. Foreign managers residing and managing at the headquarters of the company's subsidiary describe this context of the city as a religious center as something interesting, different, and in some ways exotic from their previous experience in their countries of origin. In this case, it was noted that managers from countries with a more established Christian (Catholic) tradition, for example, Italy, indicated that this context was more important to the company. *"I remember, in Italy I heard a lot about Czestochowa, I can't pronounce the name of the city very well, but a lot of Catholics in Italy heard about the Black Madonna" (interview with Italian company, product development manager, 2022)*. Thus, in an operational sense, the risk level of the religious context can be considered neutral, while the perception of the context of the place is rather positive. Figure 10.3 graphically presents the conclusions of the qualitative research conducted by means of in-depth interviews with a group of managers of foreign enterprises located in the city.

In the case of research conducted among managers of local administration and investment support agencies, an interesting tendency can be observed related to the perception of risk in the religious context of the city. At the strategic level – understood as worldwide promotional activities of the city (through national agencies and individual actions of the city) as a suitable place for foreign business location – the religious context of the city as a world pilgrimage and religious worship center is communicated as an asset and a positive distinguishing feature of the unique local brand of the place (against competing locations). *"Competition among cities in Poland to attract investment is very high, so we are trying to use our spiritual brand as a strategic differentiator. In the 1990s it seemed a rather outdated, unattractive association, but now it can build our place brand" (interview with municipal services of City of Częstochowa, manager of investor service office, 2023)*.

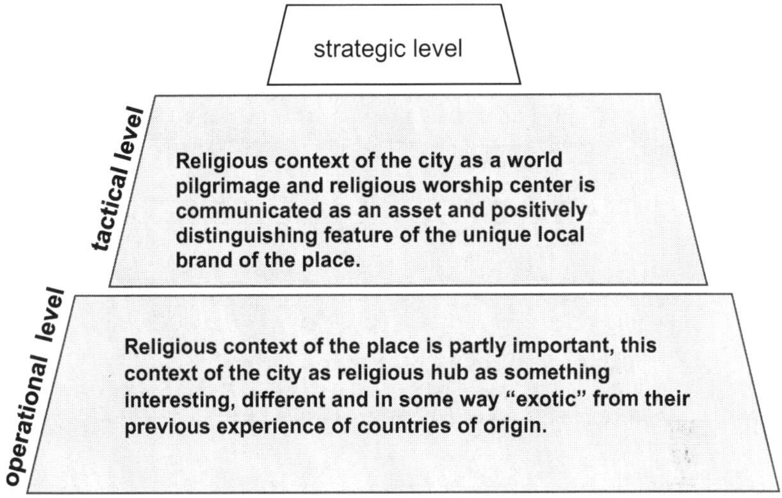

Figure 10.3 The importance of the religious context of the city by managers of international companies (author).

What is most interesting is that, at the operational level, one can observe the opposite phenomenon, which consists in recognizing that the religious context of the city is a certain risk factor associated with attracting investment. The operational level is understood as the operations and activities in a specific geographic location of the city, expansion of existing facilities, day-to-day logistics, cooperation with the community, etc. In this case, the results indicate that the religious context has no "positive" importance. Rather, it can be noted that the tendency of the city administration is to make the existing investment of foreign corporations "inviting" and communicate the non-religious functions of the city (culture, science, sports, corporate social responsibility). *"We don't want the city to be associated only with religion, after all, we have many other interesting things, sports, culture, well, and tourism in the area" (interview with Częstochowa city hall, manager at the Department of European Funds and Development, 2023)*. If, at the strategic level, we consider the religious context as an asset, the managers of the municipal administration recognize religious context at the operational level as an unfavorable factor that should be minimized. At the same time at the tactical level – understood as the concrete organizational cities' activities, consisting in preparing the place for investment, material infrastructure, issuing formal permits, and other administrative actions – the religious context has no formal and practical significance. However, it continues to be communicated in order to maintain the narrative. Figure 10.4 graphically presents the conclusions of the qualitative research on a group of city administration and agency managers.

The reported results of qualitative research are a unique case and the first example of such research carried out on the example of the city of Częstochowa.

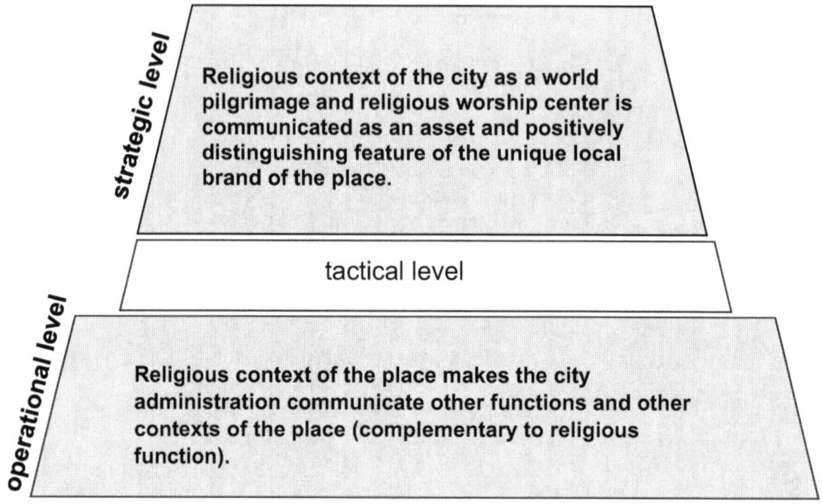

Figure 10.4 The importance of the religious context of the city, as stated by city administration and agencies (author).

Obviously, these results probably cannot be extrapolated to other contextual spaces and other decisions and risks of foreign investment, but they make an interesting contribution and seem to be an interesting input for academic discussion as well as for local development practice.

10.6 Discussion

It may seem that issues of religion have little relevance to FDI risk issues related to the process of secularization. Strulik (2012) proves the thesis that secularization is both a cause and a consequence of successful economic development. In this view, secularization is a feature of the broader process of modernization, i.e. a general change in the values associated with economic development. It is assumed that people (societies) who identify with religion place more emphasis on intangible values and struggle less for temporal goods than people (societies) who identify with secular values.

However, there is a significant body of empirical research that indicates the importance of cultural and religious distance issues in FDI localization processes. Previous research suggests that the culture and customs of FDI destination countries can determine changes in the strategic decisions of multinational corporations (Franz 2010). This seems to be particularly evident in cases where international business locations disrupt or severely interfere with the existing socio-economic system. In their study, Du, Lu, and Tao (2012) address the interplay of economic institutions and cultural distance on FDI entry using China as an example. The implication of this study is that cultural distance plays an important role in location decisions. Interesting research on global production networks, conducted by Völlers et al. (2023), proves that the relationship of business organizations operating in foreign markets with the location environment produces risk in some way. Moreover, this risk can be captured in terms of social constructivism in relation to the individual decision-making, perceptions, and expectations of actors. "[R]isk in global production networks should be seen as a driver of organizational behavior, which is socially constructed" (Völlers et al. 2023, 1848). This seems to be an approach that takes a similar research lens to the research presented in this chapter, as it focuses on human and narrative factors rather than economic mechanisms.

It seems reasonable to assume that the academic debate on the importance of the religious context within investment risk or the risk of operating in a specific investment location is rather limited. This is because current research focuses on country or national conditions, or investment risks associated with an area of religious cultures, such as Catholics, Muslims, or Protestants (Iannaccone 1998). As Díez-Esteban, Farinha, and García-Gómez (2019, 37) stressed "recent research is focused quite heavily on the socioeconomic consequences of religion, using economic theory and more sophisticated statistical tools". A more common area of research is the analysis of individual risk behavior (including organizational risk) in conjunction with religious beliefs (Shu, Sulaeman, and Yeung 2012). It follows that the vast majority of research is based on quantitative techniques and maximally large data sets; qualitative research on the specific space of a "sacred place" is particularly rare.

As Hergueux (2011) notes among the potential vectors of cultural proximity, religion has received some particular attention. At the same time, it should be noted that research on the relationship between the religious context of a place and aspects of location and business operations, including foreign investment, is not well developed. As Van Buren et al. (2020,799) state "Religion has been in general neglected or even seen as a taboo subject in organizational research and management practice". Dolansky and Alon (2008, 29), on the other hand, emphasize that "research into the links between religion and foreign direct investment is scarce".

10.7 Conclusion

Answering the research question posed at the beginning of the study: *endogenous urban strong religious context of the place can be a risk factor in the perception of foreign business investing and operating in this location?* the following conclusions can be arrived: On the basis of the qualitative research conducted and the use of three levels of analysis, i.e. strategic, tactical, and operational, it can be stated that the relationships between the perception of risk investment by foreign corporations and the presence of a strong local religious context vary. In the case of interviews with managers of foreign corporations, the relationship was observed at the meso level (tactical) understood as positive significance of the local brand of the place, which causes geographic and socio-cultural identification of the FDI location. At the micro (operational) level, the results are inconclusive, at the macro level (strategic), no relationship was observed. In the case of institutional actors, a link was observed between strategic actions (macro level) and the specificity of the shrine space, communicated as a competitive advantage. At the micro (operational) level, the sacred function of the city is downplayed in favor of other complementary urban spheres.

The research was done on a specific example of a city, a specific religious dimension of the sacrum space and specific foreign corporations, and it is not known that it can be generalized to other cases. However, it seems that the presented research can contribute to the discussion on the importance of the context space in the activities of foreign enterprises, investments in the space of the place, and the perception of the risk of FDI. Studies on the role of religious or spiritual context are rather rare, although, in an increasingly uniform world and globalized culture, they can be an interesting contribution to scientific discussions. Moreover, in Poland, for example, studying the relationship of religion and/or the Catholic Church to economic spaces or socio-territorial development is still a strong taboo. Another difficulty is the fact that, when religion functions as a historical and cultural force that affects the economy, its influence is often indirect and diffuse (van Buren et al. 2020).

In the case of directions for further research, it may be fruitful to assume that the religious context is a strong binding factor, as Hergueux (2011) noted there is a view that in countries with weak institutions, religion plays a greater role as a building block of some social and economic glue than in countries with strong institutions. Perhaps an interesting area would be to explore various phenomena of cultural, symbolic, topical, or heterotopic contexts as factors of territorial diversity, as Barro and McCleary (2003) note that the country's religious diversity facilitates FDI. One way or another, areas as regions of the world become more similar, and business practices and concepts become more and more standardized, there is always something to be discovered in the study of the diversity of intangible spaces.

References

Abbas, Z., M. Abbas, and K. Abbas. 2022. Religious freedom, FDI and Economic Growth. *Journal of Development and Social Sciences* 3 (2):1171–84. doi: 10.47205/jdss.2022(3-II)104

Act of Parliament of May 10, 2018 on the promotion of new investments, *Journal of Laws*, item 1162.

Alon, I., and J. Spitzer. 2003. Does religious freedom affect country risk assessment? *Journal of International and Area Studies* 10:51–62.

Barro, R., and R. McCleary. 2003. Religion and Economic Growth across Countries. *American Sociological Review* 68 (5):760–81.

Basco, R., and L. Suwala. 2020. Spatial Familiness – A bridge between family business and economic geography. In *A research agenda for family business. A way ahead for the field*, ed. A. Calabrò, 185–212. Cheltenham: Edward Elgar. doi: 10.4337/9781788974073.00017

Bryman, A. 2012. *Social research methods*. 4th ed. Oxford: Oxford University Press.

City of Częstochowa. Official Website. Accessed February 20, 2023. https://www.czestochowa.pl/history-of-the-city-of-czestochowa

Díez-Esteban, J. M., J. B. Farinha, and C. D. García-Gómez. 2019. Are religion and culture relevant for corporate risk-taking? International evidence. *BRQ Business Research Quarterly* 22 (1):36–55. doi: 10.1016/j.brq.2018.06.003

Dolansky, E., and I. Alon. 2008. Religious freedom, religious diversity, and Japanese foreign direct investment. *Research in International Business and Finance* 22 (1):29–39.

Drucker, P. F. 1954. *The Practice of Management*. New York: Harper and Row.

Du, J., Y. Lu, and Z. Tao. 2012. Institutions and FDI location choice: The role of cultural distances. *Journal of Asian Economics* 23:210–23.

FDI Business Financial Times. 2019. Global Free Zones of the Year 2019. Accessed 01.11.2024, https://www.ksse.com.pl/files/?id_plik=4360

Franz, M. 2010. The role of resistance in a retail production network: Protests against supermarkets in India. *Singapore Journal of Tropical Geography* 31 (3):317–29.

Fuchs, M., S. Henn, M. Franz, and R. Mudambi. 2017. Introduction. Managing Culture and Interspace in Cross-border Investments. Building a Global Company. In *Managing Culture and Interspace in Cross-border Investments. Building a Global Company*, ed. M. Fuchs, S. Henn, M. Franz, and R. Mudambi, 1–9. Abingdon: Routledge.

Gereffi, G., J. Humphrey, and T. Sturgeon. 2005. The governance of global value chains. *Review of international political economy* 12 (1):78–104.

Gereffi, G., and M. Korzeniewicz. 1994. *Commodity Chains and Global Capitalism*. Westport: Praeger Publishers.

Hall, E. 1959. *The Silent Language*. New York: Doubleday and Company.

Hall, E. 1966. *The Hidden Dimension*. New York: Doubleday and Company.

Hergueux, J. 2011. How does religion bias the allocation of Foreign Direct Investment? The role of institutions. *Économie internationale* 128:53–76.

Hodge, B. J., W. P. Anthony, and L. M. Gales. 1996. *Organization Theory: a Strategic Approach*, 5th ed. Upper Saddle River: Prentice Hall.

Hofstede, G. 1984. Cultural dimensions in management and planning. *Asia Pacific Journal of Management* 1 (2):81−99.

Hofstede, G. 2011. Dimensionalizing Cultures: The Hofstede Model in Context. *Online Readings In Psychology and Culture* 2 (1). doi:10.9707/2307–0919.1014

Hong, S., J. Lee, F. D. Oh, and D. Shin. 2023. Religion and foreign direct investment. *International Business Review* 32 (1):102035. doi: 10.1016/j.ibusrev.2022.102035

Iannaccone, L. R. 1998. Introduction to the Economics of Religion. *Journal of Economic Literature* 36 (3):1465−95.

Isard, W. 1956. *Location and space-economy*. Cambridge: MIT Press.

Jackson, J. H., and C. P. Morgan. 1982. *Organization Theory*. 2nd ed. Upper Saddle River: Prentice Hall.

Jupp, V. 2004. *The SAGE dictionary of social and cultural research methods*. London: SAGE.

Kazunobu, H., F. Kimura, and H.-H. Lee. 2013. How does country risk matter for foreign direct investment? *The Developing Economies* 51:60–78.

Kittler, M., D. Rygl, and A. Mackinnon. 2011. Beyond culture or beyond control? Reviewing the use of Hall's high-low-context concept. *International Journal of Cross Cultural Management* 11 (1):63–82.

Krugman, P. 1991. *Geography and Trade*. Cambridge: MIT Press.

La Croix International. 2020. Our Lady of Fatima welcomed 6.3 million pilgrims in 2019. La Croix International, February 10. Accessed December 12, 2022. https://international.la-croix.com/news/religion/our-lady-of-fatima-welcomed-63-million-pilgrims-in-2019/11784

Leavy, B. 1984. Assessing country risk for foreign investment decisions. *Long Range Planning* 17 (3):141–50.

Lee, J., K. Lee, and F. D. Oh. 2023. Religion and Derivative Use: Evidence from the Hedge Fund Industry. *Journal for the Scientific Study of Religion* 62 (2):451–75. doi: 10.1111/jssr.12837

López-Duarte, C., M. M. Vidal-Suárez, and B. González-Diaz. 2015. Impact of cultural positions on FDI's entry mode. *Cross Cultural Management* 22 (3):509–26.

Lösch, A. 1940. Die räumliche Ordnung der Wirtschaft. Eine Untersuchung über Standort, Wirtschaftsgebiete und internationalen Handel [The spatial organisation of the economy. A study of location, economic areas and international trade]. Jena: G. Fischer.

Lucke, N., and S. Eichler. 2016. Foreign direct investment: the role of institutional and cultural determinants. *Applied Economics* 48 (11):935–56.

Mansour, W., and M. Jlassi. 2014. The effect of religion on financial and investing decisions. In *Investor behavior: The psychology of financial planning and investing*, ed. H. K. Baker, and V. Ricciardi, 135–51. John Wiley and Sons.

Mintzberg, H. 1979. *The Structuring of Organizations: A Synthesis of the Research*. Englewood Cliffs: Prentice Hall.

Mróz, F. 2021. The Impact of COVID-19 on Pilgrimages and Religious Tourism in Europe During the First Six Months of the Pandemic. *Journal of Religion and Health* 60:625–45. doi: 10.1007/s10943-021-01201-0

Niedźwiedź, A. 2000. Obraz Matki Boskiej Częstochowskiej w kręgu mitologii narodowej, (The image of Our Lady of Częstochowa in national mythology). *Religia i polityka. Wydawnictwo Uniwersytetu Jagiellońskiego*:75–85.

Pachura, P. 2015. *Koncepcja bliskości w kształtowaniu innowacyjności przedsiębiorstw* (The concept of proximity in shaping the innovativeness of enterprises). Toruń: Dom Organizatora.

Pachura, P. 2021. *Trójprzestrzenie. Modelowanie teorii przestrzeni organizacji*. (TriSpaces. Modelling organisational space theory). Warszawa: Scholar.

Pawlicki, P. 1897. *Droga Żelazna Warszawsko – Wiedeńska w 50 letnim okresie istnienia*. (Warsaw-Vienna Iron Road in 50 years of existence). Warszawa.

Perlmutter, H. 1969. The Tortuous Evolution of the Multinational Corporation. *Columbia World Journal of Business* 1:9−18.

Perlmutter, H., and A. Heenan. 1974. How multinational should your top managers be? *Harvard Business Review* 52:121−32.

Rodríguez, N. D. 2016. Country risk in foreign direct investment: similarities and differences with country risk in exports. *Nómadas. Critical Journal of Social and Juridical Sciences* 49: 83–98. doi: 10.5209/NOMA.53538

Russell, H. M., D. Ariail, K. Taken Smith, and L. M. Smith. 2023. Religiosity and Risk: Association of Judeo-Christian Ethicality with a Sustainable Business Environment. *Journal of Risk and Financial Management* 16 (9):394. doi: 10.3390/jrfm16090394

Schlunze, R. D., M. Plattner, W. Baber, and N. Agola. 2012. *Spaces of international economy and management*. London: Palgrave Macmillan.

Shenkar, O. 2012. Cultural distance revisited: Towards a more rigorous conceptualization and measurement of cultural differences. *Journal of International Business Studies* 43:1−11.

Shu, T., J. Sulaeman, E. P. Yeung. 2012. Local Religious Beliefs and Mutual Fund Risk-Taking Behaviors. *Management Science* 58 (10):1779−96.

Sobalski, F., and M. Nowak. 2017. Krótka historia przemysłu Częstochowy XIX–XXI w. (The short history of the industry of Częstochowa). *Ziemia Częstochowska* 17:39–62.

Soussane, J., D. Mansouri, and Z. Mansouri. 2022. Does Religious Proximity Affect FDI Location Choice? An Empirical Analysis of Outward FDI from Morocco to 54 Host Countries. *Organizations and Markets in Emerging Economies* 13 2(26):426−42. doi: 10.15388/omee.2022.13.87

Strulik, H. 2012. From Worship to Worldly Pleasures: Secularization and Long-Run Economic Growth. *CRC Discussion Paper* 116. University of Goettingen.

Suwala, L. 2014. *Kreativität, Kultur und Raum – Ein wirtschaftsgeographischer Beitrag am Beispiel des kulturellen Kreativitätsprozesses* [Creativity, culture and space – An economic-geographical contribution using the example of the cultural creativity process]. Wiesbaden: Springer.

Suwala. L. 2021. Concepts of Space, Re-figuration of Spaces and Comparative Research – Perspectives from Economic Geography and Regional Economics. *Forum Qualitative Sozialforschung. Forum: Qualitative Social Research* 22 (3). doi:10.17169/fqs-22.3.3789

Suwala, L., P. Pachura, and R. D. Schlunze. 2022. Management Geography – Making Place for Space in Management Thought. *Polish Journal of Management Studies* 25 (2):323–40. doi:10.17512/pjms.2022.25.2.21

Tocar, S. 2019. Religiosity and foreign direct investment: the case of Romania. *Cross-Cultural Management Journal* 21 (1):57–66.

Van Buren, H. J., J. Syed, and R. Mir. 2020. Religion as a Macro Social Force Affecting Business: Concepts, Questions, and Future Research. *Business & Society* 59 (5):799–822. doi: 10.1177/000765031984509

Völlers, P., T. Neise, P. Verfürth, M. Franz, F. Bücken, and K. P. Schumacher. 2023. Revisiting risk in the Global Production Network approach 2.0 – Towards a performative risk narrative perspective. *Environment and Planning A: Economy and Space* 55 (8):1838–58. doi:10.1177/0308518X231169288

Weber, A. 1929. *The Theory of the Location of Industries*. Chicago: Chicago University Press.

Yavan, N. 2010. The Location Choice of Foreign Direct Investment Within Turkey: An Empirical Analysis. *European Planning Studies* 18 (10):1675–1705.

Yip, S. 1992. *Total global strategy*. Englewood Cliffs: Prentice Hall.

Index

Pages in *italics* refer to figures and pages in **bold** refer to tables

9781032601076